Innovative Composite Materials for Sound Absorption and Insulation

Innovative Composite Materials for Sound Absorption and Insulation

Editor

Francesco Martellotta

MDPI • Basel • Beijing • Wuhan • Barcelona • Belgrade • Manchester • Tokyo • Cluj • Tianjin

Editor
Francesco Martellotta
Dipartimento di Scienze
dell'Ingegneria Civile e
dell'Architettura
Politecnico di Bari
Bari
Italy

Editorial Office
MDPI
St. Alban-Anlage 66
4052 Basel, Switzerland

This is a reprint of articles from the Special Issue published online in the open access journal *Materials* (ISSN 1996-1944) (available at: www.mdpi.com/journal/materials/special_issues/mater_sound_absorpt_insul).

For citation purposes, cite each article independently as indicated on the article page online and as indicated below:

LastName, A.A.; LastName, B.B.; LastName, C.C. Article Title. *Journal Name* **Year**, *Volume Number*, Page Range.

ISBN 978-3-0365-2381-1 (Hbk)
ISBN 978-3-0365-2380-4 (PDF)

Contents

About the Editor

Francesco Martellotta

Francesco Martellotta is a Full Professor of Building Physics and Building Energy Systems at the Department of Civil Engineering and Architecture, Politecnico di Bari. His scientific interests include: architectural acoustics; sustainable energy use in buildings; characterization of sustainable building materials; thermal comfort; assessment of indoor environment quality; and noise in offices. He contributed to secure and manage several fundings within public competitive financing programs and from private companies. He is author/co-author of more than 100 scientific papers, nearly half of them published in peer-reviewed journals. In 2007, he was Keynote lecturer at the International Symposium on Room Acoustics. He serves as Associate Technical Editor of the Journal of Audio Engineering Society and as Associate Editor of the Journal of Acoustical Society of America. He is currently serving as a guest editor for several MDPI journals. He is a full member of AIA, AES, and ASA.

materials

Editorial

Innovative Composite Materials for Sound Absorption and Insulation: Where We Are and Where We Are Going

Francesco Martellotta

Dipartimento di Scienze dell'Ingegneria Civile e dell'Architettura, Politecnico di Bari, Via Orabona, 4, 70125 Bari, Italy; francesco.martellotta@poliba.it; Tel.: +39-080-5963631

Materials with sound-absorbing or sound-insulating properties have been rapidly evolving in recent years due to several reasons. On one side, there is the ever-increasing awareness of the adverse effects that noise and lack of acoustic comfort may have on human health. On the other, the availability of more sophisticated fabrication techniques, calculation methods, and new materials, has stimulated researchers and, more and more frequently, industry to develop customized materials with improved properties.

For sound-absorbing materials, combining a sufficiently wide frequency range of action with reduced mass and thickness, as well as with aesthetic requirements (due to the need to use such materials in spaces like classrooms, restaurants, train stations, etc.), is becoming essential, not to mention durability and ease of maintenance, which should always be considered. For sound-insulating materials, aesthetic problems are typically not an issue, but other aspects, such as balancing low dynamic stiffness with load-bearing properties, become important in order to ensure the best performance. In addition, whatever the purpose, such devices should be sustainable, have a low impact in terms of life cycle assessment, and possibly involve the use of recycled materials or natural products.

All of the above questions clearly represent a challenge for researchers, but at the same time, they offer new opportunities to experiment with cutting-edge solutions, like those based on the use of composite materials, as well as including nanotechnologies, "green" vegetal and animal fibers, and metal/ceramic/polymer matrixes, up to the use of acoustic metamaterials.

In this Special Issue of *Materials* [1], all of these topics are covered, providing an interesting picture of the current trends. The potential of perforated panels, that, thanks to 3D printing and refined numerically controlled manufacturing processes may now offer flexibility of use, was investigated by Li et al. [2] to obtain improved low-frequency performance thanks to a parallel arrangement of four different perforated panel absorbers in which the number and dimensions of apertures ensure a smooth manufacturing process compared to micro-perforated panels. The analytical models available to study such a group of materials also showed very good accuracy in predicting actual behavior, thus representing a further element in favor of such applications. A parallel arrangement of perforated plates with "extended" tubes was also investigated with reference to their diffuse field behavior [3]. A significant dependence of sound absorption as a function of incidence angle was found, and as a function of the period absorber layout, which made sound absorption decrease when the period was comparable with the wavelength. However, proper design ensured good sound absorption in the mid–low frequency range when using such a parallel arrangement of perforated absorbers.

The potential offered by additive manufacturing techniques in terms of customization of the porous cell structure of materials aimed at obtaining broad-band sound absorption was investigated by Cavalieri et al. [4]. They showed that, at least in theory, given the current limitations of the fabrication techniques, it could be possible to obtain a significantly high and broad-band sound absorption by using numerical optimization of geometrical

Citation: Martellotta, F. Innovative Composite Materials for Sound Absorption and Insulation: Where We Are and Where We Are Going. *Materials* **2021**, *14*, 1954. https://doi.org/10.3390/ma14081954

Received: 7 April 2021
Accepted: 9 April 2021
Published: 14 April 2021

properties of porous cells. Investigation on the role of grading properties and anisotropy showed that they can contribute to further maximizing absorption. Similarly, the role of the geometrical and microstructural properties of granular molecular sieves (i.e., zeolite crystals) was experimentally investigated by Zhou et al. [5]. They found that the intrinsic porosity of the granules is less important than the actual grain size and overall thickness of the layer in affecting sound absorption. In any case, a good fit with phenomenological model predictions was found, allowing for possible optimization of the material properties.

A second group of papers focused its attention on the use of natural or recycled components to create a skeleton of porous materials with sound-absorbing purposes. Cigarette butts of different lengths and under different conditions (smoked or non-smoked) were used by Gómez Escobar et al. [6] to produce small samples that were tested in a standing wave tube. The results showed that acceptable sound absorption could be obtained from about 2 kHz on, with small variations depending on the butts' length and smoking conditions. Rubino et al. [7] investigated the potential of recycled textile waste to produce composite sound-absorbing panels in combination with bio-based binders like chitosan and gum Arabic. Sound absorption coefficients, flow resistivity, and other physical properties were determined and the results showed relatively little influence of the binder on the acoustic behavior. Conversely, sample density and flow resistivity played a major role in changing the frequency dependence of sound absorption. The use of natural materials as the main component of granular porous sound absorbers was also the topic of research from Liuzzi et al. [8]. In this case, almond skins were dried and then mixed with different binders (including gum Arabic and polyvinyl acetate glue) to obtain, after further drying, a rigid panel. The acoustic properties in this case were enhanced by the very complex microstructure that originates from the mixture, resulting in higher absorption coefficients appearing at lower frequencies than for other absorbing materials of similar thickness. The use of environmentally friendly fire retardants was investigated by Liu and Hu [9] with reference to polyurethane foams, which are well known for their sound-absorbing properties, in combination with phase change materials.

Fediuk et al. [10] reviewed the current literature on the topic of innovative concretes with acoustic-oriented properties. Thus, treatments oriented to improve both sound-absorbing and sound-insulating characteristics were discussed, including porous and aerated concretes, concretes with special aggregates like recycled rubber crumbs, expanded polystyrene, synthetic fibers, recycled aggregates, mollusk shells, and foam glass. Detailed case studies of building components are provided.

The last two papers of the Special Issue specifically address the topic of sound insulation. Santoni et al. [11] investigated the use of wood flour as filler of wood plastic composite (WPC) panels. WPC boards are analyzed both numerically and experimentally with reference to the vibrational and transmission loss behavior of such panels. Junctions and boundary conditions proved to significantly affect the results, pointing out the importance of a proper design and modeling of such devices. Finally, Ehrig et al. [12] investigated the potential of compressible constrained layer damping (CCLD) that is a semi-active, lightweight-compatible solution for vibration mitigation based on a base structure, a constraining plate, and a compressible open-cell foam core in between, enabling the adjustment of the structure's vibration behavior by changing the core compression using different actuation pressures. The results showed that such changes affected transmission loss in the range between 1 kHz and 5 kHz, with the best performance appearing when maximum compression was applied.

Overall, a challenging and promising scenario is depicted, suggesting that in the coming years, substantial innovations might move from research labs to the market, actually contributing to improved performance combined with a lower environmental impact.

Funding: This research received no external funding.

Conflicts of Interest: The author declares no conflict of interest.

References

1. Martellotta, F. Innovative Composite Materials for Sound Absorption and Insulation. Available online: https://www.mdpi.com/journal/materials/special_issues/mater_sound_absorpt_insul (accessed on 13 April 2021).
2. Li, X.; Wu, Q.; Kang, L.; Liu, B. Design of Multiple Parallel-Arranged Perforated Panel Absorbers for Low Frequency Sound Absorption. *Materials* **2019**, *12*, 2099. [CrossRef] [PubMed]
3. Li, D.; Chang, D.; Liu, B. Diffuse Sound Absorptive Properties of Parallel-Arranged Perforated Plates with Extended Tubes and Porous Materials. *Materials* **2020**, *13*, 1091. [CrossRef] [PubMed]
4. Cavalieri, T.; Boulvert, J.; Gabard, G.; Romero-García, V.; Escouflaire, M.; Regnard, J.; Groby, J.-P. Graded and Anisotropic Porous Materials for Broadband and Angular Maximal Acoustic Absorption. *Materials* **2020**, *13*, 4605. [CrossRef] [PubMed]
5. Zhou, B.; Zhang, J.; Li, X.; Liu, B. An Investigation on the Sound Absorption Performance of Granular Molecular Sieves under Room Temperature and Pressure. *Materials* **2020**, *13*, 1936. [CrossRef] [PubMed]
6. Gómez Escobar, V.; Rey Gozalo, G.; Pérez, C.J. Variability and Performance Study of the Sound Absorption of Used Cigarette Butts. *Materials* **2019**, *12*, 2584. [CrossRef] [PubMed]
7. Rubino, C.; Bonet Aracil, M.; Gisbert-Payá, J.; Liuzzi, S.; Stefanizzi, P.; Zamorano Cantó, M.; Martellotta, F. Composite Eco-Friendly Sound Absorbing Materials Made of Recycled Textile Waste and Biopolymers. *Materials* **2019**, *12*, 4020. [CrossRef] [PubMed]
8. Liuzzi, S.; Rubino, C.; Stefanizzi, P.; Martellotta, F. Performance Characterization of Broad Band Sustainable Sound Absorbers Made of Almond Skins. *Materials* **2020**, *13*, 5474. [CrossRef] [PubMed]
9. Liu, D.; Hu, A. The Influence of Environmentally Friendly Flame Retardants on the Thermal Stability of Phase Change Polyurethane Foams. *Materials* **2020**, *13*, 520. [CrossRef] [PubMed]
10. Fediuk, R.; Amran, M.; Vatin, N.; Vasilev, Y.; Lesovik, V.; Ozbakkaloglu, T. Acoustic Properties of Innovative Concretes: A Review. *Materials* **2021**, *14*, 398. [CrossRef]
11. Santoni, A.; Bonfiglio, P.; Fausti, P.; Marescotti, C.; Mazzanti, V.; Pompoli, F. Characterization and Vibro-Acoustic Modeling of Wood Composite Panels. *Materials* **2020**, *13*, 1897. [CrossRef]
12. Ehrig, T.; Dannemann, M.; Luft, R.; Adams, C.; Modler, N.; Kostka, P. Sound Transmission Loss of a Sandwich Plate with Adjustable Core Layer Thickness. *Materials* **2020**, *13*, 4160. [CrossRef] [PubMed]

materials MDPI

Article

Design of Multiple Parallel-Arranged Perforated Panel Absorbers for Low Frequency Sound Absorption

Xin Li, Qianqian Wu, Ludi Kang and Bilong Liu *

School of Mechanical & Automobile Engineering, Qingdao University of Technology, No. 777 Jialingjiang Road, Qingdao 266520, China
* Correspondence: liubilong@qut.edu.cn; Tel.: +86-136-8337-1062

Received: 22 May 2019; Accepted: 27 June 2019; Published: 29 June 2019

check for updates

Abstract: A particular structure that consists of four parallel-arranged perforated panel absorbers (PPAs) is proposed for the low frequency sound absorption within a constraint space. The apertures of the perforated panels are set to ≥1.5 mm, and the number of orifices is much less and therefore easier to be produced in comparison with that of the micro perforated panel (MPP). A simple approximation model by using acoustic-electrical analogy is described to calculate the sound absorption coefficient of such device subject to normal wave incidence. Theoretical and experimental results demonstrate that the device can provide more than one octave sound absorption bandwidth at low frequencies.

Keywords: perforated panel; absorber array; low frequency absorption

1. Introduction

For an acoustical wave in the low frequency range, due to its large wavelength and super penetrating power, it is difficult to be handled by a thin structure. Micro perforated panel has been widely used as a wideband sound absorber replacing of traditional porous materials in circumstances requiring fireproof and environmental protection [1–3]. MPP absorber has sufficient acoustic resistance and small acoustic reactance provided by its sub-millimeter perforation, and can provide a bandwidth of one or two octaves near the resonance for sound absorption in comparison with other resonance based sound absorbers. One deficiency is that the MPP absorber requires larger cavity depth for low frequency sound absorption. Newly developed sound absorbers need to reduce the cavity depth while similar or better acoustical performance to the MPP absorber. Furthermore, although MPP absorber has tremendous potential for wide-band absorption up to 3 or 4 octaves when the aperture is set between 0.1–0.3 mm [4], the orifices are too many to be produced at low cost. For example, if a MPP absorber has aperture of 0.3 mm, the average sound absorption coefficient in the frequency range of 200–750 Hz could reach above 0.7. But an amount of 190,000 holes per square meter need to be punched on a panel with a thickness of 0.5 mm. Thus, wide application of MPP in low-frequency noise control is being hampered by high-cost manufacturing technology.

In recent years, for the purpose of expanding the absorption bandwidth, a series of extended structures based on MPP in parallel or series have been presented to introduce additional resonances. The combined structure of double or multi-layered MPPs and air cavities arranged in tandem along the direction of sound wave propagation were proposed successively [2,5–8]. The results show that the absorption bandwidth is extended to lower frequencies due to the additional resonance peak, while at the cost of increased air cavity. An alternately approach to extend the sound absorption bandwidth is to introduce multi-resonance through a compound MPP absorber array. Zha et al. [3] designed a parallel structure of the MPPs with two different cavities and measured its normal incidence absorption coefficients by impedance tube. They also explained that the appearance of two different resonance peaks is due to different cavity depths. Sakagami et al. [9] later proposed a combination of two different MPP absorbers for obtaining a wideband absorption device, such as the same MPPs with different air-cavity depths or different MPPs with the same air-cavity depth. Its absorption characteristics by considering the excess attenuation caused by the impedance discontinuity was described. The physical absorption mechanisms of an MPP absorber array consisting of three MPP absorbers with different cavity depths was investigated by Wang et al. [10]. They concluded that strong local resonance occurs due to different reactance matching conditions for the component MPP absorbers and the resonance frequencies shift due to inter-resonator interactions. Moreover, the oblique incidence sound absorption of parallel arrangement of four MPP absorbers in diffuse field was also investigated both numerically and experimentally [11]. Although the parallel MPP absorber array have the potential to improve sound absorption bandwidth due to multi-resonance of component MPPs, the reported perforation aperture is limited to 0.3–0.8 mm, and therefore the cost for the massive processing will be extremely high.

An elongated tube arrays of flexible tube bundles [12–15] or rigid extended tubes [16,17] attached to the perforated panel or MPP were designed to improve the low frequency sound absorption in a limited space. Among them, a perforated panel by parallel-arranged extended tubes (PPET) designed by Li et al. achieved more than one octave absorption bandwidth at low frequencies in a constrained space of 100mm. When the diameter of extend tubes were set at 2.9–5.1 mm and the length of extend tubes were 20–51 mm, the average sound absorption coefficient in the range of 120–250 Hz was more than 0.7. Although the perforation apertures of PPETs are much larger and the holes numbers are much less in comparison with that of the MPP absorbers, the lattice arrangement of a bundle of extend tubes makes the structure more complex and difficult to be manufactured.

In this paper, a composite structure of parallel-arranged perforated panel absorbers (PPAs) with unequal apertures of 1.5–4 mm is designed to improve low-frequency sound absorption in a limited space. The sound absorption performance of the designed absorber is almost equal to that of PPET in [16], and it is more practical because this device is simply a perforated panel of different apertures backed with same air cavity, without the extended tube bundle. As follows, a theoretical model of a particular structure consisting of four parallel-arranged perforated panel absorbers is described in Section 2, and the optimization parameters are given in in Section 3. After that, the test verification is shown in Section 4 and the conclusion is in Section 5.

2. Theoretical Model

2.1. Acoustic Impedance of the Perforated Panel

A perforated panel is regarded as a parallel connection of numerous millimeter-level tubes with a certain depth. Therefore, the acoustic impedance of the perforated panel is equal to the acoustic impedance of a single tube divided by its perforation ratio. At first, Rayleigh studied the sound wave propagation in the tube, Crandall later simplified the sound wave propagation for short tubes [18]. When the diameter and length of the tube are far less than the wavelength of sound wave, the specific acoustic impedance of the single tube is defined as

$$z = \frac{\Delta P}{\overline{u}} = j\omega\rho t\left[1 - \frac{2}{k\sqrt{-j}}\frac{J_1(k\sqrt{-j})}{J_0(k\sqrt{-j})}\right]^{-1} \tag{1}$$

where ΔP is the sound pressure difference between the two ends of the tube, $\overline{u}$ is the average velocity across the cross-section of the tube, $k = d/2\sqrt{\omega\rho/\eta}$ is the ratio of the inner radius to the viscous boundary layer thickness inside the tube, $\omega = 2\pi f$ is the angular frequency, ρ is the air density, d and t are the diameter and length of the tube.

For Equation (1), Maa [1] further gave the approximate formula of the acoustic impedance of the tube for all k values, which is expressed as

$$z = \frac{32\eta t}{d^2}\left(1+\frac{k^2}{32}\right)^{1/2} + j\omega\rho t\left(1+\left(9+\frac{x^2}{2}\right)^{-1/2}\right) \tag{2}$$

For this design, the perforated panel of a certain thickness is assumed to be acoustically rigid, and the vibration effect of the plate under acoustic load is ignored. In addition, due to the acoustic radiation at both ends of the tube and the air flow friction on the surface plate, the correction of the acoustic resistance and the acoustic mass of the tube should be added to Equation (2). Therefore, the normalized characteristic impedance for the perforated panel with the perforation ratio p is written as

$$Z_{PP} = \frac{32\eta t}{p\rho c d^2}\left(\sqrt{1+\frac{k^2}{32}} + \frac{\sqrt{2}}{32}k\frac{d}{t}\right) + \frac{\omega t}{pc}\left(1 + 1\Big/\sqrt{3^2+\frac{k^2}{2}} + 0.85\frac{d}{t}\right) \tag{3}$$

2.2. Acoustic Impedance of the Parallel-Arranged Perforated Panel Absorber

The perforated panel absorber array consists of four sub-PPA arranged in parallel, its structure diagram is shown in Figure 1a. Using the electrical equivalent circuit model in Figure 1b, this perforated panel absorber is regarded as a parallel combination of four RLC branch. Moreover, the back cavity is divided into four sub-cavities by the rigid clapboards to avoid the interaction between the sub-perforated panel resonators. Then, the normalized acoustic impedance for each sub-perforated panel absorber is expressed as

$$Z_i = r_i + j\omega m_i - j\cot(\omega D_i/c) \tag{4}$$

$$r_i = \frac{32\eta t_i}{p_i \rho c {d_i}^2} k_{r_i}, k_{r_i} = \sqrt{1+\frac{{k_i}^2}{32}} + \frac{\sqrt{2}}{32}k_i\frac{d_i}{t_i} \tag{5}$$

$$\omega m_i = \frac{\omega t_i}{p i c} k_{m_i}, k_{m_i} = 1 + \left(3^2+\frac{{k_i}^2}{2}\right)^{-1/2} + 0.85\frac{d_i}{t_i} \tag{6}$$

The total acoustic impedance of the four parallel-arranged PPAs is written as

$$Z = \left(\sum_{i=1}^{4}\frac{\phi_i}{Z_i}\right)^{-1} \tag{7}$$

where the subscript $i = 1, 2, 3, 4$ denotes the number of sub-perforated panel, p_i is the perforation ratio of each sub-perforated panel, η is the viscous coefficient of the air, c is the speed of sound wave in air and ρc is the characteristic impedance in air, r_i and m_i are the normalized acoustic resistance and reactance of sub-perforated panel respectively, t_i is the thickness of each sub-perforated panel, D_i is the depth of the back cavity of each sub-perforated panel.

Then, for the normal incidence condition, the sound absorption coefficient of the parallel-arranged PPAs is expressed as

$$\alpha = \frac{4\mathrm{Re}al(Z)}{(1+\mathrm{Re}al(Z))^2 + (\mathrm{Im}ag(Z))^2} \tag{8}$$

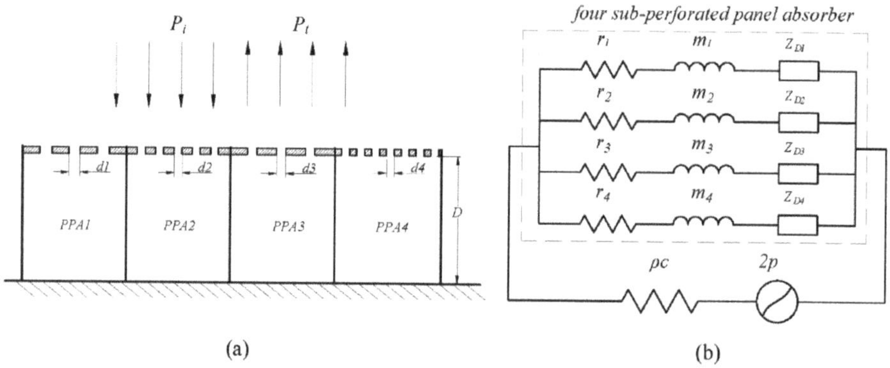

Figure 1. Schematic diagram and electrical equivalent circuit model: (**a**) Schematic of the four parallel-arranged perforated panel absorbers. (**b**) The electrical equivalent circuit model of the four parallel-arranged perforated panel absorbers.

3. Model Optimization and Comparison

Simulated annealing is an effective, fast and straightforward general probabilistic algorithm, which is widely used to obtain the best possible configuration for sound absorption systems due to its global optimization [19,20]. Similarly, in this paper, the parameters of sound absorption structure are optimized by simulated annealing method to seek the maximum average sound absorption within a specific frequency range.

3.1. Comparison of Sound Absorption between Perforated Panel and MPP in Low Frequencies

In the frequency range of 100–300 Hz, the sound absorption of single perforated panel (d = 0.7 mm) and MPP (d = 1.5 mm) with the same ratio (t/d = 1.3) are plotted in Figure 2. By contrast, it can be seen that sound absorption coefficient of the perforated panel is significantly higher than that of a single MPP (d = 0.7 mm), and the resonance peak moves to the lower frequency. The perforated panel has a higher sound absorption coefficient because the acoustic resistance of the perforated panel matches the characteristic impedance of air by reasonably adjusting the ratio of the larger aperture and plate thickness. From Figure 2b, it can be confirmed that the normalized acoustic resistance of the MPP is much higher than 1 because of the overdamping caused by the tiny aperture, while that of the perforated panel is close to 1 in the range of 100–200 Hz. In addition, Figure 2c shows that the normalized zero acoustic reactance of the perforated panel occurs at 185 Hz, and the normalized zero acoustic reactance of MPP occurs at 220 Hz. This intuitively explains that the resonance frequency of the perforated panel is lower than that of MPP. In other words, the shift of resonance to lower frequency is mainly attributed to the increase of acoustic mass with the increase of aperture depth.

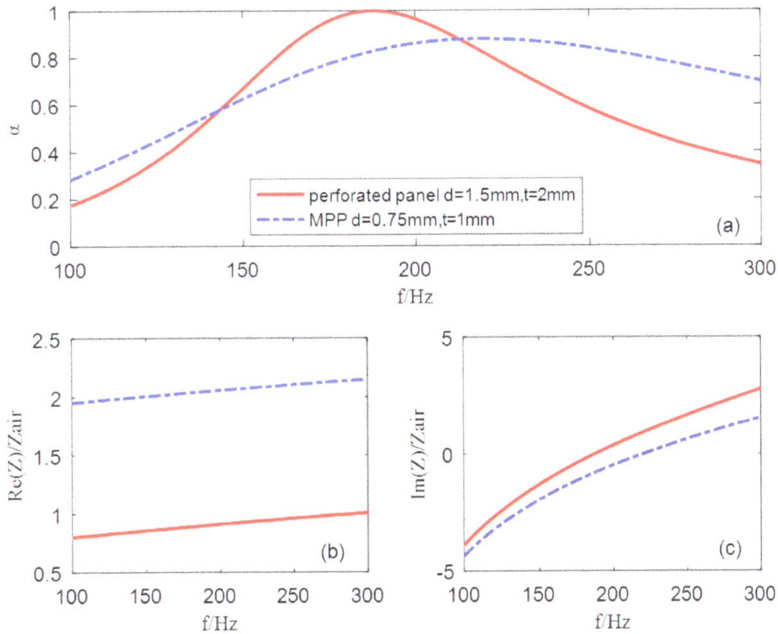

Figure 2. Comparison of the sound absorption of single perforated panel and MPP: (**a**) sound absorption coefficient; (**b**) normalized acoustic resistance; (**c**) normalized acoustic reactance. The parameters of the single optimized perforated panel are as follows: d = 1.5 mm, t = 2 mm, D = 100 mm and p = 0.46%; the parameters of the single optimized MPP are as follows: d = 0.75 mm, t = 1 mm, D = 100 mm and p = 0.33%.

However, an obvious defect is that the effective sound absorption bandwidth of single perforated panel is narrow. Therefore, a combined structure of multiple perforated panels in parallel needs to be designed and optimized to expand the sound absorption bandwidth for the low frequencies.

3.2. Comparison of Sound Absorption between PPAs and Other Resonant Structures in the Same Depth Cavity

In order to compare the sound absorption performance of the parallel-arranged PPAs and the existing resonant absorption structures in the same frequency range, the maximum average sound absorption coefficient in the range of 120–250 Hz was selected for comparison. Table 1 lists optimized parameters of the parallel-arranged PPAs, PPETs [16] and MPP absorber. The sound absorption coefficients of the three comparative structures are plotted in Figure 3.

Table 1. Optimization parameters of three sound absorption structures.

Parameter	PPAs				PPETs [16]				MPP
	PPA_1	PPA_2	PPA_3	PPA_4	$PPET_1$	$PPET_2$	$PPET_3$	$PPET_4$	
d (mm)	3.1	2.8	2.2	2	5.1	3.5	3.1	2.9	0.8
p (%)	0.32	0.46	0.62	0.93	3.26	3.74	5.31	4.43	0.41
t (mm)	2.5	2.5	2.5	2.5	51	40	40	20	2
D (mm)	100	100	100	100	100	100	100	100	100

Due to the nature of its own Helmholtz resonator, the absorption bandwidth of MPP occurs near the resonance, while the absorption deviating from the resonance drops rapidly. The parallel-arranged PPAs extends the absorption bandwidth through four resonances. For the parallel-arranged PPAs, four resonances and three anti-resonances occur due to each sub-PPA as a resonator and its coupling with each other. As shown in Figure 3, the four resonant frequencies occur at 130 Hz, 160 Hz, 190 Hz, and 230 Hz, and their sound absorption coefficients are 0.87, 0.96, 0.96, and 0.96 respectively. Moreover, the effective sound absorption bandwidth of the parallel-arranged PPAs is about 130 Hz (the sound absorption coefficients of 121–251 Hz are higher than 0.7). In the same restricted space of D = 100 mm, the sound absorption of the parallel-arranged PPAs is significantly better than that of a single MPP in the frequency ranges of 110–165 Hz and 215–250 Hz. That is, in comparison with that of a single MPP, the improvement of the bandwidth is about 55 Hz under 165 Hz and about 45 Hz above 215 Hz. Therefore, this proposed structure can obviously extend the sound absorption bandwidth compared with a single MPP in the same frequency range.

In the range of 120–250 Hz, the sound absorption performance of the parallel-arranged PPAs is close to that of PPETs absorber [16]. Although the sound absorption coefficients of the first resonance and anti-resonances are slightly lower than that of PPETs, they are greater than 0.7, especially the transition between anti-resonance and resonance becomes flatter as the frequency tends to be higher. Besides, by comparing geometric parameters, the aperture of the parallel-arranged PPAs is between 2–3.1 mm, and that of PPETs is between 2.9–5.1 mm. Even though the aperture of the parallel-arranged PPAs is smaller than that of PPETs, its depth diameter ratio is much lower than that of the latter. Considering the processing technology, the hole with smaller depth diameter ratio is more convenient to be processed by common machining. Also, the perforation rate of the perforated panel absorber is lower than that of the PPETs. As for the parallel-arranged PPAs, about 57 holes per square decimeter are drilled on a panel; while for the PPETs, about 193 thin hollow tubes of different lengths per square decimeter are attached to a perforated panel in parallel. So, the designed structure is not complicated and simple to manufacture.

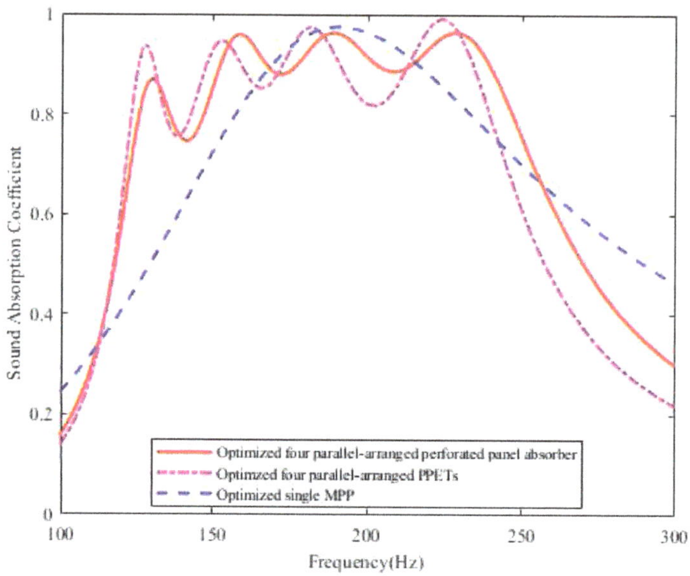

Figure 3. Comparison of the sound absorption of the four parallel-arranged PPAs and PPETs, MPP absorber in the frequency range of 120–250 Hz.

3.3. Comparison of PPAs with Different Cavity Depth

Figure 4 shows the sound absorption optimization of the four parallel-arranged PPAs with different cavity depth. The four solid lines (from left to right) represent the optimized sound absorption of four different cavities of 100 mm, 70 mm, 50 mm and 35 mm, corresponding to four different frequency bands, such as 200–450 Hz, 300–550 Hz, 400–650 Hz, and 400–750 Hz respectively. And the sound absorption coefficient of each peak and peak valley is above 0.85. The effective absorption bandwidth is about 3–4 1/3 octaves. Besides, the apertures of the PPAs ≥ 1.5 mm and the maximum thickness of the plate is 2.5 mm, as listed in Table 2. Therefore, when the aperture is designed to be ≥1.5 mm, this four parallel-arranged PPAs structure can be applied to different frequency bands for sound absorption at low frequencies by adjusting the depth of the back cavity and optimizing structural parameters.

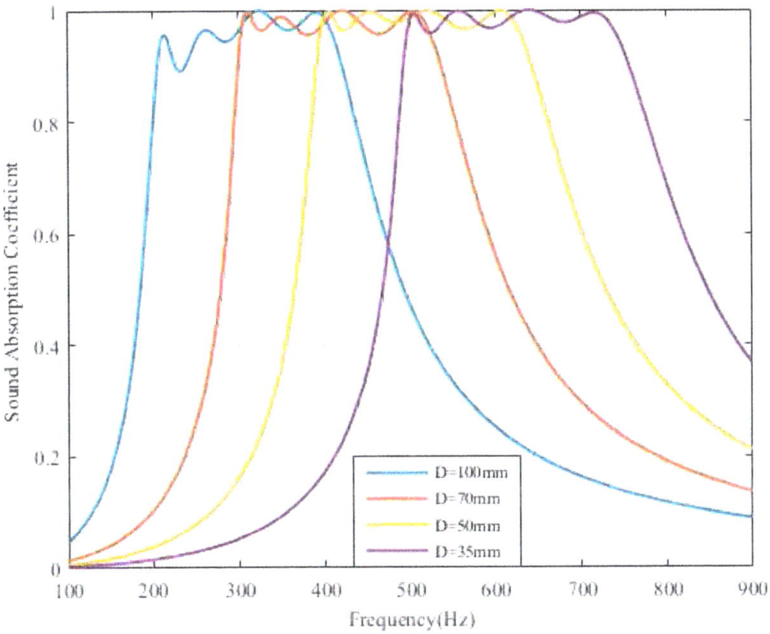

Figure 4. Optimization of the parallel-arranged PPAs with different cavity depth.

Table 2. Optimization parameters of PPAs with different cavity depth.

Parameter	D = 100 mm				D = 70 mm			
	PPA_1	PPA_2	PPA_3	PPA_4	PPA_1	PPA_2	PPA_3	PPA_4
d (mm)	2.4	1.5	1.5	1.6	3.5	1.5	1.6	1.5
p (%)	0.74	2.94	1.63	1.01	1.20	1.79	1.18	3
t (mm)	2.3	2.3	2.3	2.3	2.1	2.1	2.1	2.1
Parameter	**D = 50 mm**				**D = 35 mm**			
	PPA_1	PPA_2	PPA_3	PPA_4	PPA_1	PPA_2	PPA_3	PPA_4
d (mm)	1.7	3.1	1.5	1.5	1.8	1.5	1.5	4
p (%)	1.44	1.39	3	1.98	1.66	3	2.19	1.81
t (mm)	2.2	2.2	2.2	2.2	2.5	2.5	2.5	2.5

4. Experimental Validation

The proposed structure can be made of metallic materials (stainless steel, aluminum) or non-metallic materials such as plastic wood and etc. The aperture of the PPAs is not less than 1.5mm and the maximum depth of the hole is 2.5 mm. For stainless steel or aluminum, mechanical or laser drilling can be used for bulk processing. The focus of this research is on the configuration and implementation of the structure, thus, a perforated plate made of resin produced by 3D technology is considered for verification.

The measurements are carried out in an impedance tube (SW422) according to ISO 10534-2 [21], and the instruments used in the experiment are shown in Figure 5. Based on the transfer function method, this device measures the sound pressure at two locations using two microphones and calculates the normal incident sound absorption coefficient. When two microphones connected to the data acquisition instrument are installed at position 0 and position 2 of the impedance tube, the sound absorption coefficient of 63–500 Hz is measured; when two microphones are placed in position 1 and position 2, the absorption coefficient of 400–1600 Hz is measured. The samples in the experiment are composed of perforated panel, clapboard and cylindrical shell, and an integrated structure of resin materials produced by 3D technology. The thickness of clapboard and cylindrical shell is 2mm and 2 mm, respectively. Besides, the sample must be sealed and mounted at the back of the impedance tube to ensure the presence of multiple resonance peaks.

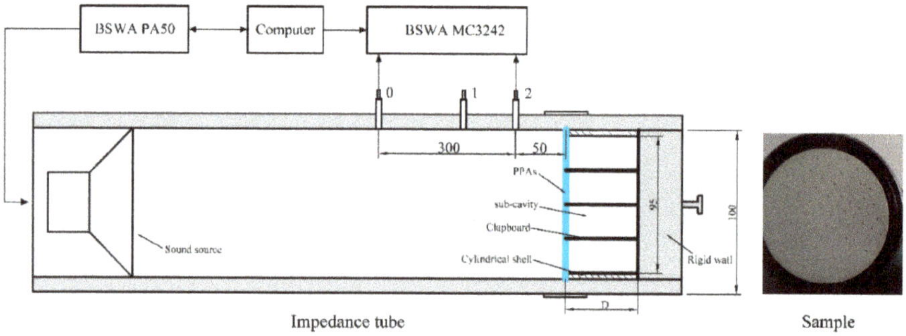

Figure 5. The experimental Setup.

In order to illustrate the sound absorption characteristics of the perforated panel absorber in different frequency range, two structures with different parameters are made for experimental verification. Sample A with aperture of 2.1mm, 2.5mm,3 mm is used for 120–300 Hz, and sample B with aperture of 1.5 mm,1.6 mm and 2.2 mm is used for 150–450 Hz.The parameters of samples are listed in Tables 3 and 4, respectively.

As plotted in Figures 6 and 7, it can be seen that the theoretical prediction is in a relatively good agreement with the experimental results. The measurement results in Figure 6 show that sample A has an effective sound absorption bandwidth of 140 Hz within 135–275 Hz, and this compact structure can absorb large wavelength when the thickness is only 1/27 of the incident wavelength. Besides, the measured absorption coefficient of sample B exceeds 0.6 from 155–455 Hz, and the effective absorption bandwidth is extended to about four 1/3 octaves, as shown in Figure 7. Therefore, it is further confirmed that this designed perforated panel absorber has great potential in extending the sound absorption bandwidth at low frequency.

Table 3. Parameters of the four parallel-arranged PPAs for 120–250 Hz.

Sample A	d (mm)	p (%)	t (mm)	D (mm)
PPA_1	3	0.36	2.5	100
PPA_2	2.5	0.50	2.5	100
PPA_3	2.1	0.64	2.5	100
PPA_4	2.1	0.96	2.5	100

Table 4. Parameters of the four parallel-arranged perforated panel absorbers for 150–450 Hz.

Sample B	d (mm)	p (%)	t (mm)	D (mm)
PPA_1	2.2	0.39	2	100
PPA_2	1.6	0.63	2	100
PPA_3	1.6	2.52	2	100
PPA_4	1.5	1.26	2	100

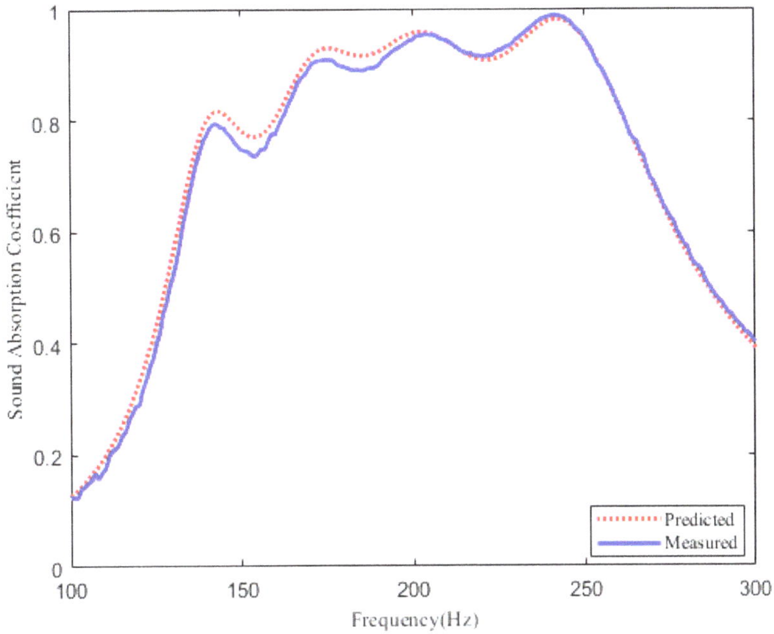

Figure 6. Sound absorption coefficient of the four parallel-arranged PPAs for 125–300 Hz. Red dotted line: predicted results. Blue solid line: measured results.

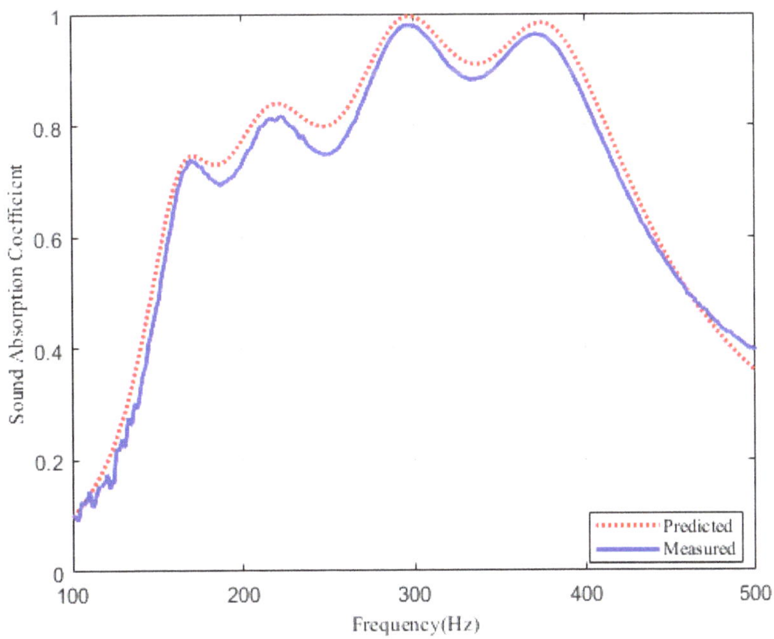

Figure 7. Sound absorption coefficient of the four parallel-arranged PPAs for 150–450 Hz. Red dotted line: predicted results. Blue solid line: measured results.

5. Conclusions

A parallel-arranged perforated panel absorber for low frequency sound absorption is investigated. A theoretical model is described to predict the acoustic properties. The sound absorption performance of the parallel-arranged PPAs with that of the existing resonant structure is discussed, which provide the basis to illustrate the feasibility of the designed structure. Samples for two frequency bands of 120–250 Hz and 150–450 Hz are designed and measured. The effective sound absorption covers the frequency range from 135 Hz to 275 Hz and 155–455 Hz respectively. Theoretical and experimental results prove that a wideband sound absorption of more than one octave can be achieved by a parallel combination of perforated panels with same cavity depth at low frequencies. Further work is to study the sound absorption performance of perforated panels with different cavity depth for low-frequency acoustic absorption and the perforated panel array in diffuse field.

Author Contributions: For this research, X.L. designed and conducted the experiments, and finished the manuscript. Q.W. and L.K. provided some useful suggestion in the experiments and manuscript editing. B.L. gave technical guidance and supervision in this design.

Funding: This study was co-supported by Taishan Scholar Program of Shandong (No. ts201712054), Shandong Provincial Natural Science Foundation, China (No. ZR2018BEE020) and A Project of Shandong Province Higher Educational Science and Technology Program (No. J18KA018).

Conflicts of Interest: The authors declare no conflict of interest.

References

1. Maa, D.Y. Theory and Design of Microperforated Panel Sound-Absorbing Constructions. *Sci. Sin.* **1975**, *18*, 55–71.
2. Maa, D.Y. Wide-Band Sound Absorber Based on Microperforated Panels. *Chin. J. Acoust.* **1985**, *4*, 3–14.
3. Cha, X.; Jian, K.; Zhang, T.; Zhou, X.; Fuchs, H. Application Approach for Microperforated Panel Sound Absorbers. *Acta Acust.* **1994**, *19*, 256–265.
4. Maa, D.Y. Potential of Microperforated Panel Absorber. *J. Acoust. Soc. Am.* **1998**, *104*, 2861–2866. [CrossRef]
5. Maa, D.Y. Microperforated-Panel Wideband Absorbers. *Noise Control Eng. J.* **1987**, *29*, 77–84. [CrossRef]
6. Sakagami, K.; Morimoto, M.; Koike, W. A Numerical Study of Double-Leaf Microperforated Panel Absorbers. *Appl. Acoust.* **2006**, *67*, 609–619. [CrossRef]
7. Sakagami, K.; Matsutani, K.; Morimoto, M. Sound Absorption of a Double-Leaf Micro-Perforated Panel with an Air-Back Cavity and a Rigid-Back Wall: Detailed Analysis with a Helmholtz-Kirchhoff Integral Formulation. *Appl. Acoust.* **2009**, *71*, 411–417. [CrossRef]
8. Sakagami, K.; Yairi, M.; Morimoto, M. Multiple-Leaf Sound Absorbers with Microperforated Panels: An Overview. *Acoust. Aust.* **2010**, *38*, 76–81.
9. Sakagami, K.; Nagayama, Y.; Morimoto, M.; Yairi, M. Pilot Study on Wideband Sound Absorber Obtained by Combination of Two Different Microperforated Panel (Mpp) Absorbers. *Acoust. Sci. Technol.* **2009**, *30*, 154–156. [CrossRef]
10. Wang, C.; Huang, L. On the Acoustic Properties of Parallel Arrangement of Multiple Micro-Perforated Panel Absorbers with Different Cavity Depths. *J. Acoust. Soc. Am.* **2011**, *130*, 208. [CrossRef] [PubMed]
11. Wang, C.; Huang, L.; Zhang, Y. Oblique Incidence Sound Absorption of Parallel Arrangement of Multiple Micro-Perforated Panel Absorbers in a Periodic Pattern. *J. Sound Vib.* **2014**, *333*, 6828–6842. [CrossRef]
12. Lu, Y.; Tang, H.; Tian, J.; Li, H. The Perforated Panel Resonator with Flexible Tube Bundle. *Acoust. Soc. Am. J.* **2006**, *119*, 3249. [CrossRef]
13. Lu, Y.; Tang, H.; Tian, J.; Li, H.; Yang, J. The Perforated Panel Resonator with Flexible Tube Bundles and Its Sound Absorption Measurements. In *INTER-NOISE and NOISE-CON Congress and Conference Proceedings*; Institute of Noise Control Engineering: Chicago, IL, USA, 2001; pp. 4220–4226.
14. Simon, F. Low Frequency Sound Absorption of Resonators with Flexible Tubes. *J. Acoust. Soc. Am.* **2013**, *133*, 3265. [CrossRef]
15. Simon, F. Long Elastic Open Neck Acoustic Resonator for Low Frequency Absorption. *J. Sound Vib.* **2018**, *421*, 1–16. [CrossRef]

16. Li, D.; Chang, D.; Liu, B. Enhancing the Low Frequency Sound Absorption of a Perforated Panel by Parallel-Arranged Extended Tubes. *Appl. Acoust.* **2016**, *102*, 126–132. [CrossRef]
17. Li, D.; Chang, D.; Liu, B. Enhanced Low- to Mid-Frequency Sound Absorption Using Parallel-Arranged Perforated Plates with Extended Tubes and Porous Material. *Appl. Acoust.* **2017**, *127*, 316–323. [CrossRef]
18. Crandall, I.B. *Theory of Vibrating Systems and Sound*, 2nd ed.; D. Van Nostrand Company: New York, NY, USA, 1927; p. 544.
19. Chiu, M.C.; Chang, Y.C.; Yeh, L.J.; Lan, T.S. Optimization of Perforated Double-Layer Absorbers Using Simulated Annealing. *J. Mar. Sci. Technol.* **2007**, *15*, 351–359.
20. Ruiz, H.; Cobo, P.; Jacobsen, F. Optimization of Multiple-Layer Microperforated Panels by Simulated Annealing. *Appl. Acoust.* **2011**, *72*, 772–776. [CrossRef]
21. Chung, J.Y.; Blaser, D.A. Transfer Function Method of Measuring in-Duct Acoustic Properties. Ii. Experiment. *J. Acoust. Soc. Am.* **1998**, *68*, 914–921. [CrossRef]

materials

MDPI

Article

Diffuse Sound Absorptive Properties of Parallel-Arranged Perforated Plates with Extended Tubes and Porous Materials

Dengke Li [1,*], Daoqing Chang [2,*] and Bilong Liu [3]

1 The State Key Laboratory of Heavy Duty AC Drive Electric Locomotive Systems Integration, Zhuzhou 412001, China
2 Key Laboratory of Noise and Vibration Research, Institute of Acoustics, Chinese Academy of Sciences, Beijing 100190, China
3 School of Mechanical and Automotive Engineering, Qingdao University of Technology, No. 777 Jialingjiang Road, Qingdao 266520, China; Liubilong@qut.edu.cn
* Correspondence: lidengkeyouxiang@126.com (D.L.); changdq@mail.ioa.ac.cn (D.C.)

Received: 11 February 2020; Accepted: 27 February 2020; Published: 1 March 2020

check for updates

Abstract: The diffuse sound absorption was investigated theoretically and experimentally for a periodically arranged sound absorber composed of perforated plates with extended tubes (PPETs) and porous materials. The calculation formulae related to the boundary condition are derived for the periodic absorbers, and then the equations are solved numerically. The influences of the incidence and azimuthal angle, and the period of absorber arrangement are investigated on the sound absorption. The sound-absorption coefficients are tested in a standard reverberation room for a periodic absorber composed of units of three parallel-arranged PPETs and porous material. The measured 1/3-octave band sound-absorption coefficients agree well with the theoretical prediction. Both theoretical and measured results suggest that the periodic PPET absorbers have good sound-absorption performance in the low- to mid-frequency range in diffuse field.

Keywords: sound absorption; perforated plates with extended tubes; porous materials; periodic absorber

1. Introduction

Micro-perforated panels (MPP) are widely used in the engineering noise control of vehicles, buildings, and ventilation facilities, because of their good sound-absorption performance and many advantages, such as environmental friendliness, durability, and hygienic properties. In 1975, Maa [1] proposed the idea that panels with sub-millimeter perforations could provide sufficient acoustic resistance to achieve a high absorption coefficient. At the same time, he founded the theory of MPP and presented some engineering designs. Although the MPP sound absorber has a larger frequency band of absorption compared to conventional perforated panels, it does not satisfy some engineering requirements, especially in the low frequency range. Maa [2] further introduced a double-layer MPP to increase the bandwidth of sound absorption.

In 1992, micro-perforated acrylic–glass plates were successfully used to attenuate the reverberation sound in German Bundestag Hall [3]. Later many researchers tried to improve the sound-absorptive properties of MPP, using various methods. When the impedance of the pores is comparable with the impedance of the panel or membrane on which the pores go through, their coupling effect should be considered. Kang et al. [4] presented a theoretical method to predict the sound-absorption coefficient of such structure. With appropriate parameters, the absorption coefficient can exceed 0.4 over 3–4 octaves. Lee et al. [5] theoretically studied the sound absorption of finite flexible micro-perforated panel, results show that the additional sound-absorption peak of the panel vibration effect can widen the absorption bandwidth of MPP when the resonant frequency of the flexible panel is higher than that of Helmholtz-resonance frequency of the MPP. Sakagami et al. [6] also used an electro-acoustical equivalent circuit model to study the relationship between the panel-type absorbers and MPP absorbers. Further research by Toyoda et al. [7] revealed that only the panel-type absorption caused by the eigen-mode vibration can occur independently from the Helmholtz-resonance absorption of MPP, and that the panel-type absorption caused by a mass-spring resonance cannot be utilized to widen the absorption bandwidth. Sakagami et al. [8,9] proposed a double-leaf MPP spatial sound absorber (DLMPP) which is composed of two MPP leaves, without the backing structure. A triple layer absorber composed of permeable membrane (PM) [10] and the double-leaf MPP (DLMPP) was proposed to improve the absorption performance, and the peak absorption at resonance was found to become significantly higher. The effect of a honeycomb in the air cavity was also discussed on the proposed absorber: The resonance peak is shifted to lower frequencies, and the level of absorption becomes higher at mid and high frequencies, owing to the effect of the honeycomb. Through an experiment and numerical simulation, Liu et al. [11] demonstrated that partitioning the adjoining cavity enhances the attenuation of the acoustic modes propagating transverse to the MPP. This effect was particularly noticeable at low frequencies where the acoustic response is resonant in nature. Wang et al. [12] used parallel-arranged MPPs with different air cavities to improve the bandwidth of sound absorption. Gai [13] combined L-shaped cavity with MPP to enhance the low frequency sound-absorptive performance. Tang et al. [14] found that small perforations on the face-sheet and honeycomb corrugation could improve the sound absorption at low frequencies. Huang et al. [15] investigated the effect of incompletely partitioned cavities on the sound absorption of MPP and found that if the insertion period and the length of separators were arranged appropriately, the low-frequency sound-absorption performance could be much improved. Pfretzschner et al. [16] presented a new strategy, which was named "MIU", to increase the sound-absorption bandwidth. The MIU is composed of one thick plate with large perforations and another thin plate with a very high perforation ratio. This combination could produce sound absorption over two or three octave bands.

Substantial works have focused on introducing additional Helmholtz resonators [17–19] or mechanical vibration [20,21] coupled with the MPP to improve its low-frequency sound-absorption performance. The combination of electromechanical system [22], shunted circuit loudspeaker [23], or aluminum-electrode PVDF piezoelectric film [24] with the MPP is also presented to improve the sound-absorption performance, especially in the low frequency range.

Another alternative approach for improving the low-frequency sound-absorption performance of MPP in a limited space is increasing the depth of apertures, using attached tubes [25–30]. Among them, Li et al. proposed multiple perforated plates with extended tubes (PPETs) with the MPP or porous material to obtain broadband sound absorption in the low- to mid-frequency range in a constrained space.

Some authors also investigated the sound-absorptive performance of perforated plates or micro-perforated plates periodically aligned in diffuse fields [31,32]. These studies showed that the sound-absorption performance under oblique incidence could differ from that under normal incidence. Moreover, the design guideline for diffuse sound absorption for low- to mid-frequency range is also a major concern in engineer noise control. Hence, the sound-absorption performance of PPETs with porous materials in a diffuse field requires further investigation, and in this paper, the diffuse sound absorption is theoretically and experimentally analyzed for absorbers composed of periodically arranged PPETs and porous materials.

2. Theoretical Model

A two-dimensional absorber composed of three parallel-arranged PPETs with porous material is considered. Figure 1 shows the basic module of a two-dimensional periodic absorber: G_1, G_2 and G_3 are the surface admittances of three PPETs and G_4 is the surface admittance of porous material. A plane wave, p_i impinges on the absorber at the pitch angle of θ and azimuthal angle of β. The cavity wall and the rigid backing of the PPET (as shown in Figure 2) are considered to be acoustically rigid. In this study, the effect of the vibration of perforated panel is not considered for the sake of simplicity. The period of the composite absorber in both x and y directions is represented by T, as shown in Figure 1. The tube lengths, inner diameters and cavity depths of the three PPETs are 3.3, 10 and 100 mm, respectively, and the perforations of PPET1–3 are set as 0.90%, 1.54% and 2.59%, respectively.

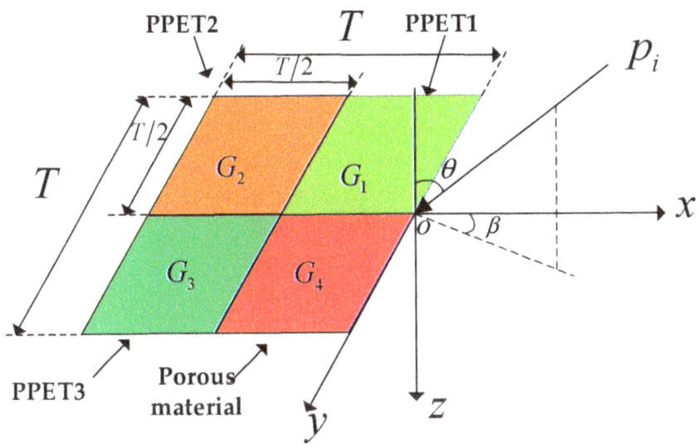

Figure 1. Schematic diagram of parallel-arranged perforated plates with extended tubes (PPETs). The angles of the incidence and azimuthal sound are defined as θ and β in this schematic.

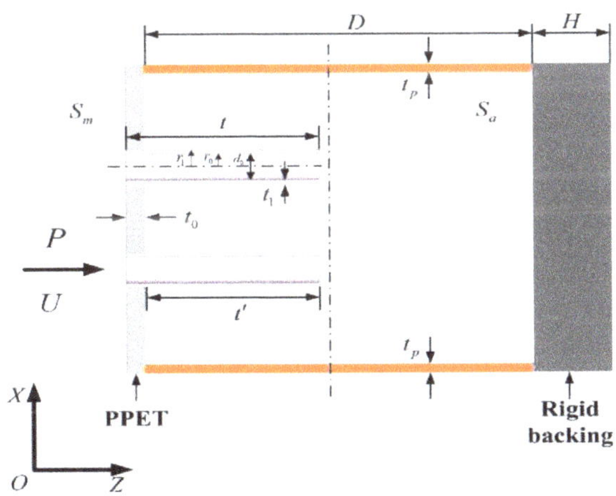

Figure 2. One unit of two-dimensional periodic absorber composed of three parallel-arranged PPETs with porous material.

2.1. Surface Impedance of the Perforated Plate with Extended Tubes

One unit of two-dimensional periodic PPET absorber are plotted in Figure 2. In the schematic diagram, S_m denotes to the cross-sectional area of this unit; t and d_0 respectively correspond to the length and inner diameter of the extended tubes; r_0 and r_1 are respectively the inner and outer radius of the extended tubes; and t_p and t_0 respectively correspond to the thickness of the cavity wall and the perforated plate. $S_0 = \pi r_0{}^2$ corresponds to the inner cross-sectional area of the extended tubes, S_a denotes the effective cross-sectional area of the back cavity, and $S_1 = \pi r_1{}^2$ corresponds to the outer cross-sectional area of the extended tube. The specific acoustic impedance was derived in Reference [27] of a PPET which has a cavity with the thickness of D.

$$Z_p = \frac{Z}{\sigma_p \rho c} = r_p + j\omega m_p - j/(\delta \tan(\omega(D - t + t_0)/c) + (\delta - \sigma') \tan(\omega(t - t_0)/c)) \quad (1)$$

$$r_p = \frac{32\eta t}{\sigma_p \rho c {d_0}^2}\left(\left(1+\frac{k^2}{32}\right)^{1/2} + \frac{\sqrt{2}kd_0}{64t}\right) \tag{2}$$

$$\omega m_p = \frac{\omega t}{\sigma_p c}\left(1+\left(9+\frac{k^2}{2}\right)^{-1/2} + 0.85\frac{d_0}{t}\right) \tag{3}$$

where η denotes the viscosity of the air; ω denotes to the angular velocity; and ρ and c are the density and the sound speed in the air, respectively. "$k = d_0\sqrt{\omega\rho/4\eta}$" corresponds to the perforation constant of the PPET; "$\sigma_p = NS_0/S_m$" denotes the perforation ratio of the PPET; "$\delta = S_a/S_m$" denotes the expansion ratio of cross-sectional area from the back cavity to the PPET; and "$\sigma' = NS_1/S_m$" denotes the ratio of the outer cross-sectional area of the extended tubes over that of the PPET (N is the number of extended tubes).

2.2. Surface Impedance of Porous Material

In the present study, an equivalent fluid model known as JCAL model is used to predict the equivalent density and modulus of the porous materials, and then the equivalent density $\rho_{eq}(\omega)$ and the equivalent modulus $K_{eq}(\omega)$ of the porous fluid are given as follows:

$$\rho_{eq}(\omega) = \frac{\alpha_\infty \rho_0}{f}\left[1 - j\frac{\sigma f}{\omega \rho_0 \alpha_\infty}\sqrt{1+\frac{4{\alpha_\infty}^2\eta\rho_0\omega}{\sigma^2\Lambda^2 f^2}}\right] \tag{4}$$

$$K_{eq}(\omega) = \frac{\gamma P_0/f}{\gamma - (\gamma-1)\left[1 - j\frac{f\kappa}{k_0' C_p \rho_0 \omega}\sqrt{1 + j\frac{4{k_0'}^2 C_p \rho_0 \omega}{\kappa {\Lambda'}^2 f^2}}\right]} \tag{5}$$

where P_0 is the ambient mean pressure; ϕ is the porosity, ω is the angular frequency; ρ_0 is the viscosity of the fluid; "$v = \eta/\rho_0 = P_r v'$" is the kinematic viscosity of the fluid; P_r is the Prandtl number of the air; and γ is the specific heat ratio.

The wave number, k_s, and characteristic impedance, Z_s, of the equivalent fluid medium are calculated as follows:

$$Z_s = \sqrt{\rho_{eq}(\omega)K_{eq}(\omega)} \tag{6}$$

$$k_s = \omega\sqrt{\rho_{eq}(\omega)/K_{eq}(\omega)} \tag{7}$$

When the angle of the sound impinges on the porous materials is θ, the surface impedance of the porous material with the thickness, L, is given by the following:

$$Z = -j\frac{Z_s k_s}{\rho c k_{sz}}\cot(k_{sz}L) \tag{8}$$

where $k_{sz} = \sqrt{{k_s}^2 - {k_x}^2 - {k_y}^2} = \sqrt{{k_s}^2 - k^2\sin^2(\theta)}$, ρc and k are the characteristic impedance and wave number of the air.

Then, an indirect characterization approach is used to obtain the transport parameters of melamine foam samples as described by Panneton and Olny [33,34], and a direct characterization of the viscous static permeability is carried out via measuring the resistivity [35]. The viscous-inertial frequency response function $\rho_{eq}(\omega)$ and the thermal frequency response function $K_{eq}(\omega)$ are measured with the three-microphone impedance tube method [36,37]. When ϕ, k_0, $\rho_{eq}(\omega)$ and $K_{eq}(\omega)$ are known, α_∞, k_0', Λ and Λ' can be experimentally estimated through an analytical inversion based on the models proposed by Johnson et al. [38] and Lafarge et al. [39]. Parameters of the porous foams are listed in Table 1.

Table 1. Parameters for the Basotect G+ and TG foam.

Parameters	σ (N·m^{-4}·s)	ϕ	α_∞	Λ (μm)	Λ' (μm)	k'_0($\times 10^{-10}$ m^2)
Basotect G+	10934 ± 182	0.994	1.04 ± 0.03	92 ± 5	197 ± 9	27 ± 1
Basotect TG	7800 ± 200	0.993	1.03 ± 0.02	134 ± 16	317 ± 32	47 ± 4

2.3. Prediction of the Diffuse-Field Sound-Absorption Properties of Periodic Absorber

First, the total sound pressure, including the incident sound wave and the scattered sound waves, is given as follows:

$$p(x,0) = p_i(x,0) + p_r(x,0) \tag{9}$$

The incident sound wave is calculated as follows:

$$p_i(x,z) = p_0 e^{j(-xk_x - yk_y + zk_z)} \tag{10}$$

where $k_{x,m}$, $k_{y,n}$ and γ_{mn} are the wavenumbers in the x, y and z directions, respectively. According to reference [32], the scattered sound wave is expressed in terms of modal expansion.

$$p_r = \sum_{m=-\infty}^{+\infty} \sum_{n=-\infty}^{+\infty} A_{mn} e^{j(-xk_{x,m} - yk_{y,n} - z\gamma_{mn})} \tag{11}$$

where A_{mn} is the unknown amplitude of the (m,n)th mode of the scattered wave. Since the scattered field is periodic in the x and y directions, the wave numbers of the (m,n)th mode of the scattered wave are as follows:

$$\begin{aligned} k_{x,m} &= k \sin\theta \cos\beta + 2m\pi/T \\ k_{y,n} &= k \sin\theta \sin\beta + 2n\pi/T \\ \gamma_{mn} &= -jk\sqrt{(\sin\theta\cos\beta + m\lambda/T)^2 + (\sin\theta\sin\beta + n\lambda/T)^2 - 1} \end{aligned} \tag{12}$$

where $\lambda = 2\pi/k$ is the wavelength. The corresponding (m,n)th acoustic mode which could propagate to the far field must satisfy the following relationship:

$$(\sin\theta\cos\beta + m\lambda/T)^2 + (\sin\theta\sin\beta + n\lambda/T)^2 \le 1 \tag{13}$$

Then, according to the relation between the sound pressure and particle velocity on the surface, we get the following:

$$\rho c v(x,y,0) = -G(x,y)p(x,y,0) \tag{14}$$

where

$$G(x,y) = \begin{cases} G_1 = \frac{1}{Z_{p1}}, \ 0 \le x < \frac{T}{2}, \ 0 \le y < \frac{T}{2} \\ G_2 = \frac{1}{Z_{p2}}, \ \frac{T}{2} \le x < T, \ 0 \le y < \frac{T}{2} \\ G_3 = \frac{1}{Z_{p3}}, \ 0 \le x < \frac{T}{2}, \ \frac{T}{2} \le y < T \\ G_4 = \frac{1}{Z_{psam}}, \ \frac{T}{2} \le x < T, \ \frac{T}{2} \le y < T \end{cases} \tag{15}$$

Inserting Equations (9) and (15) into the boundary condition Equation (14), multiplying both sides by a factor of $e^{jm(2\pi/T)x} e^{jn(2\pi/T)y}$ and then integrating the equation with respect to x and y over the period T yields a set of linear algebraic equations:

$$\begin{aligned} &\sum_{m=-\infty}^{+\infty} \sum_{n=-\infty}^{+\infty} A_{mn}\left(g_{(m-m',n-n')} + \delta_{(m-m',n-n')}\left(\frac{\gamma_{mn}}{k_0}\right)\right) = P_0\left(\delta_{m,0}\delta_{n,0}\cos\theta - g_{(m,n)}\right) \\ &m = -\infty, \cdots, +\infty \quad \& \quad n = -\infty, \cdots, +\infty \end{aligned} \tag{16}$$

where $g_{(m,n)} = \frac{1}{L_x L_y} \int_0^{L_y} \int_0^{L_x} G(x,y) e^{jm(2\pi/T)x} e^{jn(2\pi/T)y} dxdy$ and the Kronecker symbol, $\delta_{m,n}$, is defined as follows:

$$\delta_{(m-m',n-n')} = \begin{cases} 1, & m-m' = n-n' \\ 0, & m-m' \neq n-n' \end{cases} \tag{17}$$

This infinitely large system of equations will be terminated at the index limits "$m,n = \pm 2^*N$", where N is the number of elements in one period, and in this study, $N = 2$. By solving the above equations, the coefficients A_{mn} can be obtained. According to Mechel [40], the angle-dependent sound-absorption coefficient $\alpha(\theta, \beta)$ is given by Equation (18):

$$\alpha(\theta,\beta) = 1 - \left|\frac{A_{00}}{P_0}\right|^2 - \frac{1}{\cos\theta} \times \sum_{m\neq 0}\sum_{n\neq 0} \left|\frac{A_{mn}}{P_0}\right|^2 \sqrt{1 - \left(\sin\theta\cos\beta + m\frac{\lambda}{T}\right)^2 - \left(\sin\theta\sin\beta + n\frac{\lambda}{T}\right)^2} \tag{18}$$

where the summation runs over all the radiating harmonics only, the second term is the specular reflection and the third term is the scattered sound. The averaged diffuse-field sound-absorption coefficient is defined as follows [32,41]:

$$\alpha_s = \frac{1}{2\pi} \int_0^{2\pi} \left(\int_0^{\pi} \alpha_{(\theta,\beta)} \sin(2\theta) d\theta \right) d\beta \tag{19}$$

When the period width $T << \lambda$, the specular reflection is only non-evanescent reflection. In this case, the sound-absorption coefficient and normalized impedance of the periodic absorber can be derived as follows:

$$\alpha(\theta,\beta) = 1 - \left|\frac{A_{00}}{P_0}\right|^2 \tag{20}$$

$$Z(\theta,\beta) = \frac{1 - \frac{A_{00}}{P_0}}{1 + \frac{A_{00}}{P_0}} \tag{21}$$

3. Oblique-Incidence Sound-Absorption Properties of Periodic Absorber

Figure 3a shows the sound-absorption coefficients of a periodic absorber composed of units of three parallel-arranged PPETs and porous material at different incident angles (θ), when T = 10 cm, β = 0°, and the frequency interval in the calculation is 10 Hz. In the following discussion, the porous material chosen in the simulation is the Basotect TG foam (BASF(China) Co. Ltd., Beijing, China). The period length, T, is much smaller than the wavelength of the incident wave, so the sound-absorption coefficient and the normalized impedance can be calculated by using Equations (20) and (21), respectively. Three absorption peaks observed at 160, 210 and 275 Hz are owning to the Helmholtz resonances of the PPETs, while the fourth peak, observed at 800 Hz, is the resonance frequency provided by the porous material.

The influences of the incident angle, θ, on the specific resistance and reactance are shown in Figure 3b,c, respectively. It is observed that the incidence angle can greatly influence the sound-absorption coefficients of four parallel-arranged PPETs. The sound-absorption coefficient gets larger when the incidence angle is increased from 0° to 60°; it then decreases when the incidence angle goes from 60° to 89°, and it drops greatly at a near-grazing incidence angle (θ = 89°). Figure 3b shows that, at a larger incidence angle, the characteristic resistance and the reactance (absolute value) of the proposed absorber are smaller. It is clear that, when the incidence angle, θ, is equal to 0°, the specific resistance is much larger than 1, and the sound-absorption coefficient is reduced. When the incidence angle is θ = 60°, the normalized acoustic impedance matches well with the characteristic impedance of the air, and the sound-absorption coefficient is largely increased. As shown in Figure 3b, when incidence angle is θ = 89°, the normalized acoustic resistance is too small compared with the characteristic impedance of the air, so the sound-absorption coefficient is smaller. Hence, for larger sound absorption at oblique incidence, the normalized resistance of the proposed absorber should be larger than 1.

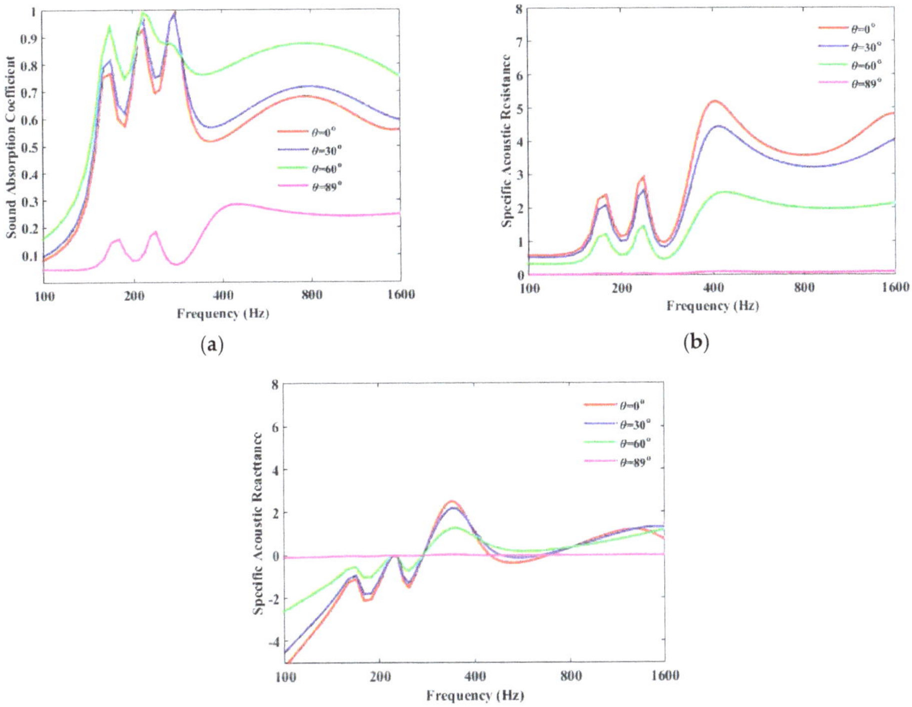

Figure 3. Comparison of the sound-absorption coefficients of the unit composed of three parallel-arranged PPETs and porous material at different angles when T = 10 cm, β = 0°. (**a**) Sound-absorption coefficient; (**b**) specific acoustic resistance; (**c**) specific acoustic reactance.

Figure 4 shows the variation of period length on the sound-absorption coefficient of the periodic absorber. When the period length, T (10 cm), is much smaller than the wavelength of the incident wave, the parallel sound absorption of the periodic absorber could be maintained. When the period length of the absorber is increased to 40 cm, the absorption coefficient drops abruptly at 840 Hz, corresponding to which the wavelength is comparable to the period length, T. When the period length of the absorber is increased to 80 cm (which is comparable to the wavelength corresponding to 420 Hz), the absorption coefficient drops abruptly at around 420 Hz. It could be concluded that the period of the proposed periodic absorber is a critical factor that controls the mechanism of parallel absorption. Hence, in a diffuse sound field, the period length of the periodic absorber should not be larger than the wavelength of interest.

The effect of the sound-absorption coefficient of the periodic absorber with the azimuthal angle β is illustrated in Figure 5. It is observed that the azimuthal angle has much less influence on the diffuse sound absorption compared with that of the incidence angle. When the period length, T, is 20 cm, the sound-absorption coefficient varies a little with the azimuthal angle in the high frequency range and remains the same in the low frequency range. When the period length, T, is 40 cm, the variation of the sound-absorption coefficient with azimuthal angle is more significant in the high frequency range.

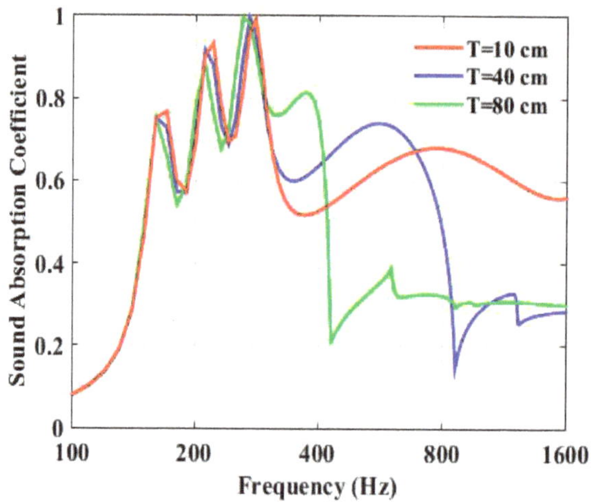

Figure 4. Influence of the period length on the sound-absorption coefficient of periodic absorber, when $\theta = \beta = 0°$.

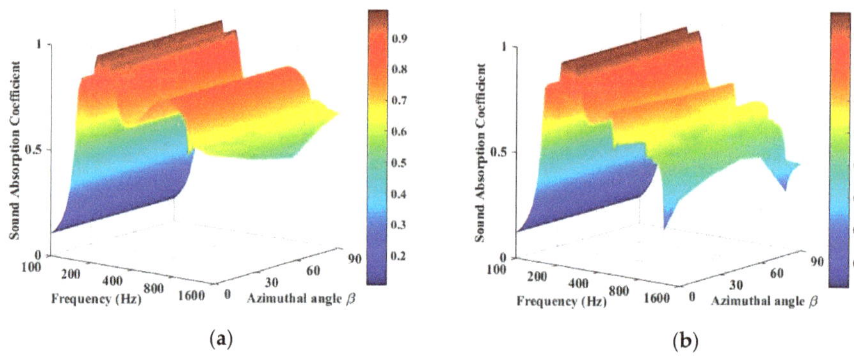

Figure 5. Sound-absorption coefficients of periodic absorber at different azimuthal angles. (**a**) T = 20 cm, $\theta = 45°$. (**b**) T = 40 cm, $\theta = 45°$.

4. Experimental Validation

The diffuse-field sound-absorption coefficient of parallel-arranged PPETs and porous material (BASF(China) Co. Ltd., Beijing, China) is measured in a reverberation room, and the measured results are compared with the theoretical predictions. The measurements are conducted in compliance with the ISO 354-2003 standard [42]. The three dimensions of the cubic chamber are 6.86, 4.94 and 5.79 m, respectively, and the total area of the test specimen is 10.3 m^2. Several curved sheets as reflectors are hung in the room. The excitation speaker is a spherical sound source. Figure 6 shows the photo picture of test sample, and the period length of the absorber, T, is 80 cm, the cavity depth of the parallel-arranged PPETs is 7 cm and the thickness of the porous material is 10 cm. The perforated panels used in the experiments were made of plastic plate, and the extended tubes were made of copper. Two kinds of Basotect foam, Basotect G+ and Basotect TG, were chosen for the experiments, and the parameters are listed in Table 1.

The measured and predicted 1/3 octave sound-absorption coefficient curves of three parallel-arranged PPETs and porous foam are shown in Figures 7 and 8. The sound-absorption peaks of PPETs are weakened due to 1/3 octave resolution. A reasonable agreement could be found in most frequency band, except below 200 Hz. The discrepancy mainly comes from the too-low modal density of the reverberation chamber in low frequency. The PPETs combined with porous material in a limited thickness of 10 cm have a good sound-absorption performance between 160 and 3150 Hz.

Figure 6. Picture of test sample.

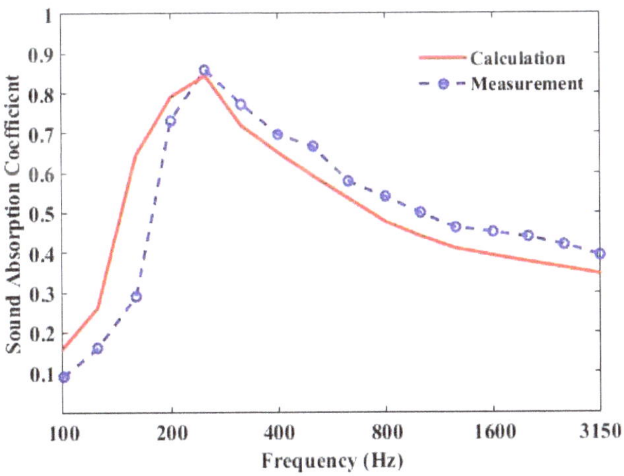

Figure 7. Comparison of the measured and calculated sound-absorption coefficient for periodic absorber composed of units of three parallel-arranged PPETs and Basotect G+ foam.

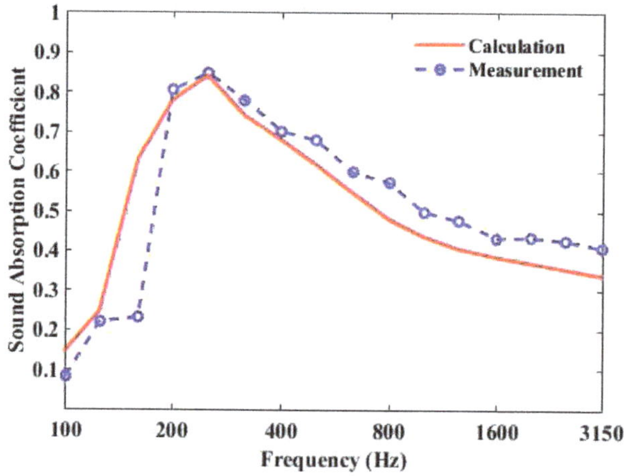

Figure 8. Comparison of the measured and calculated sound-absorption coefficient of periodic absorber composed of units of three parallel-arranged PPETs and Basotect TG foam.

5. Conclusions

The low-frequency sound-absorptive properties in diffuse field are predicted and measured for the periodically arranged sound absorbers composed of perforated plates with extended tubes (PPETs) and porous materials. The results show that increasing the incidence angle will decrease the characteristic resistance of the periodic absorber, and the sound-absorption coefficient reaches the lowest near the grazing incidence angle. The effect of the azimuthal angle on the sound-absorption coefficient is found to be insignificant. It is noted that the period of the proposed absorber is a critical factor that controls the mechanism of parallel absorption, and when the period of absorber arrangement is comparable to the wavelength of incident wave, the sound-absorption coefficient drops abruptly. Both theoretical and measured results show that periodically arranged PPETs combined with porous material could keep good sound-absorptive performance in low- to mid- frequency in diffuse sound field.

Author Contributions: For research, D.L. performed the theoretical analysis and designed the experiments and finished the manuscript. D.C. gave guidance and supervision in this design. B.L. provided useful suggestions in the experiments and manuscript editing. All authors have read and agreed to the published version of the manuscript.

Funding: This work was partly supported by the project of State Grid Corporation of China: Audible noise characteristics of electrical equipment and noise control on the boundary of converter stations in ±800 kV HVDC transmission systems and the National Natural Science Foundation of China (Grant No. 11374326).

Conflicts of Interest: The authors declare no conflict of interest.

References

1. Maa, D.Y. Theory and design of micro-perforated sound absorbing constructions. *Sci. Sin.* **1975**, *18*, 55–71.
2. Maa, D.Y. Microperforated-panel wideband absorbers. *Noise Control. Eng. J.* **1987**, *29*, 77–84. [CrossRef]
3. Fuchs, H.V.; Zha, X. Acrylic-glass sound absorbes in the Plenum of the Deutscher Bundestag. *Appl. Acoust.* **1997**, *51*, 211–217. [CrossRef]
4. Kang, J.; Fuchs, H.V. Predicting the absorption of open weave textiles and micro-perforated membranes backed by an air space. *J. Sound Vib.* **1999**, *220*, 905–920. [CrossRef]
5. Lee, Y.Y.; Lee, E.W.M.; Ng, C.F. Sound absorption of a finite flexible micro-perforated panel backed by an air cavity. *J. Sound Vib.* **2005**, *287*, 227–243. [CrossRef]
6. Sakagami, K.; Morimoto, M.; Yairi, M. A note on the relationship between the sound absorption by microperforated panels and panel/membrane-type absorbers. *Appl. Acoust.* **2009**, *70*, 1131–1136. [CrossRef]

7. Toyoda, M.; Mu, R.L.; Takahashi, D. Relationship between Helmholtz-resonance absorption and panel-type absorption in finite flexible microperforated-panel absorbers. *Appl. Acoust.* **2010**, *71*, 315–320. [CrossRef]
8. Sakagami, K.; Morimoto, M.; Koike, W. A numerical study of double-leaf microperforated panel absorbers. *Appl. Acoust.* **2006**, *67*, 609–619. [CrossRef]
9. Sakagami, K.; Nakamori, T.; Morimoto, M.; Yairi, M. Double-leaf microperforated panel space absorbers: A revised theory and analysis. *Appl. Acoust.* **2009**, *70*, 703–709. [CrossRef]
10. Sakagami, K.; Fukutani, Y.; Yairi, M.; Morimoto, M. A theoretical study on the effect of a permeable membrane in the air cavity of a double-leaf microperforated panel space sound absorber. *Appl. Acoust.* **2014**, *79*, 104–109. [CrossRef]
11. Liu, J.; Herrin, D.W. Enhancing micro-perforated panel attenuation by partitioning the adjoining cavity. *Appl. Acoust.* **2010**, *17*, 120–127. [CrossRef]
12. Wang, C.; Huang, L. On the acoustic properties of parallel arrangement of multiple micro-perforated panels with different cavity depths. *J. Acoust. Soc. Am.* **2010**, *130*, 208–218. [CrossRef] [PubMed]
13. Gai, X.; Xing, T.; Li, X.; Zhang, B.; Wang, F.; Cai, Z.N.; Han, Y. Sound absorption of microperforated panel with L shape division cavity structure. *Appl. Acoust.* **2017**, *122*, 41–50. [CrossRef]
14. Tang, Y.; Li, F.; Xin, F.; Lu, T. Heterogeneously perforated honeycomb-corrugation hybrid sandwich panel as sound absorber. *Mater. Des.* **2017**, *134*, 502–512. [CrossRef]
15. Huang, S.; Li, S.; Wang, X.; Mao, D. Micro-perforated absorbers with incompletely partitioned cavities. *Appl. Acoust.* **2017**, *126*, 114–119. [CrossRef]
16. Pfretzschner, J.; Cobo, P.; Simòn, F.; Cuesta, M.; Fernández, A. Microperforated insertion units: An alternative strategy to design microperforated panels. *Appl. Acoust.* **2006**, *67*, 62–73. [CrossRef]
17. Park, S.H. Acoustic properties of micro-perforated panel absorbers backed by Helmholtz resonators for the improvement of low-frequency sound absorption. *J. Sound Vib.* **2013**, *332*, 4895–4911. [CrossRef]
18. Zhao, X.; Yu, Y.; Wu, Y. Improving low-frequency sound absorption of micro-perforated panel absorbers by using mechanical impedance plate combined with Helmholtz resonators. *Appl. Acoust.* **2016**, *114*, 92–98. [CrossRef]
19. Gai, X.; Xing, T.; Li, X.; Zhang, B.; Wang, W. Sound absorption of microperforated panel mounted with helmholtz resonators. *Appl. Acoust.* **2016**, *114*, 260–265. [CrossRef]
20. Zhao, X.; Fan, X. Enhancing low frequency sound absorption of micro-perforated panel absorbers by using mechanical impedance plates. *Appl. Acoust.* **2015**, *88*, 123–128. [CrossRef]
21. Gai, X.; Xing, T.; Li, X.; Zhang, B.; Cai, Z.N.; Wang, F. Sound absorption of microperforated panel with membrane cell and mass blocks composite structure. *Appl. Acoust.* **2018**, *137*, 98–107. [CrossRef]
22. Chang, D.; Liu, B.; Li, X. An electromechanical low frequency panel sound absorber. *J. Acoust. Soc. Am.* **2010**, *128*, 639–645. [CrossRef] [PubMed]
23. Tao, J.; Jing, R.; Qiu, X. Sound absorption of a finite micro-perforated panel backed by a shunted loudspeaker. *J. Acoust. Soc. Am.* **2014**, *135*, 231–238. [CrossRef] [PubMed]
24. Duan, X.; Wang, H.; Li, Z.; Zhu, L.; Chen, R.; Kong, D.; Zhao, Z. Sound absorption of a flexible micro-perforated panel absorber based on PVDF piezoelectric film. *Appl. Acoust.* **2015**, *88*, 84–89. [CrossRef]
25. Lu, Y.; Li, X.; Tian, J.; Wei, W. The perforated panel resonator with flexible tube bundle and its acoustical measurements. In Proceedings of the Inter-Noise 2001, The Hague, The Netherlands, 27–30 August 2001.
26. Yahya, I.; Harjana, H. Sound absorption improvement strategy for QRD element. In Proceedings of the 20th International Congress on Sound & Vibration 2013, Bangkok, Thailand, 7–11 July 2013.
27. Li, D.; Chang, D.; Liu, B. Enhancing the low frequency sound absorption of a perforated panel by parallel-arranged extended tubes. *Appl. Acoust.* **2016**, *102*, 126–132. [CrossRef]
28. Li, D.; Chang, D.; Liu, B. Enhanced low- to mid-frequency sound absorption using parallel-arranged perforated plates with extended tubes and porous material. *Appl. Acoust.* **2017**, *127*, 316–323. [CrossRef]
29. Simon, F. Long Elastic Open Neck Acoustic Resonator for low frequency absorption. *J. Sound Vib.* **2018**, *421*, 1–16. [CrossRef]
30. Simon, F. Long Elastic Open Neck Acoustic Resonator in flow. In Proceedings of the Inter-Noise 2016, Hamburg, German, 21–24 August 2016.
31. Takahashi, D. A new method for predicting the sound absorption of perforated absorber systems. *Appl. Acoust.* **1997**, *51*, 71–84. [CrossRef]

32. Wang, C.; Huang, L.; Zhang, Y. Oblique incidence sound absorption of parallel arrangement of multiple micro-perforated panel absorbers in a periodic pattern. *J. Sound Vib.* **2014**, *333*, 6828–6842. [CrossRef]
33. Panneton, R.; Olny, X. Acoustical determination of the governing viscous dissipation in porous media. *J. Acoust. Soc. Am.* **2006**, *119*, 2027–2040. [CrossRef]
34. Olny, X.; Panneton, R. Acoustical determination of the governing thermal dissipation in porous media. *J. Acoust. Soc. Am.* **2008**, *123*, 814–824. [CrossRef] [PubMed]
35. Stinson, R.; Daigle, G. Electronic system for the measurement of flow resistance. *J. Acoust. Soc. Am.* **1988**, *83*, 2422–2428. [CrossRef]
36. Iwase, T.; Izumi, Y.; Kawabata, R. A new measuring method for sound propagation by using sound tube without any air spaces. In Proceedings of the Inter-Noise and Noise-Congress and Conference Proceedings, Christchurch, New Zealand, 16–18 November 1998.
37. Salissou, Y.; Panneton, R. Wideband characterization of the complex wave and characteristic impedance of sound absorbers. *J. Acoust. Soc. Am.* **2010**, *128*, 2868–2876. [CrossRef]
38. Johnson, D.L.; Koplik, J.; Dashen, R. Theory of dynamic permeability and tortuosity in fluid-saturated porous media. *J. Fluid Mech.* **1987**, *176*, 379–402. [CrossRef]
39. Lafarge, D.; Lemarinier, P.; Allard, J.F.; Tarnow, V. Dynamic compressibility of air in porous structures and audible frequencies. *J. Acoust. Soc. Am.* **1997**, *102*, 1995–2006. [CrossRef]
40. Mechel, F.P. Sound Fields at Periodic Absorbers. *J. Sound Vib.* **1990**, *136*, 379–412. [CrossRef]
41. Drotleff, H.; Wack, R.; Leistner, P. Absorption of periodically aligned absorber strips in concrete structures. *Build. Acoust.* **2009**, *16*, 233–256. [CrossRef]
42. *ISO 354:2003, Acoustics-Measurement of Sound Absorption in a Reverberation Room*; ISO: Geneva, Switzerland, 2003.

materials

MDPI

Article

Graded and Anisotropic Porous Materials for Broadband and Angular Maximal Acoustic Absorption

Théo Cavalieri [1,2,*], Jean Boulvert [1,2,3], Gwénaël Gabard [1], Vicent Romero-García [1], Marie Escouflaire [2], Josselin Regnard [2] and Jean-Philippe Groby [1]

1 Laboratoire d'Acoustique de l'Université du Mans, LAUM-UMR CNRS 6613, Le Mans Université, Avenue Olivier Messiaen, 72085 Le Mans CEDEX 9, France; jean.boulvert@univ-lemans.fr (J.B.); gwenael.gabard@univ-lemans.fr (G.G.); vicente.romero@univ-lemans.fr (V.R.-G.); jpgroby@univ-lemans.fr (J.-P.G.)

2 Safran Aircraft Engines, Rond Point René Ravaud - Réau, 77550 Moissy-Cramayel, France; marie.escouflaire@safrangroup.com (M.E.); josselin.regnard@safrangroup.com (J.R.)

3 Laboratoire d'Analyse Vibratoire et Acoustique, LAVA, Department of Mechanical Engineering, École Polytechnique de Montréal, P.O. Box 6079 Station Centre-ville, Montréal, QC H3C 3A7, Canada

* Correspondence: theo.cavalieri@univ-lemans.fr

Received: 1 September 2020; Accepted: 30 September 2020; Published: 16 October 2020

check for updates

Abstract: The design of graded and anisotropic materials has been of significant interest, especially for sound absorption purposes. Together with the rise of additive manufacturing techniques, new possibilities are emerging from engineered porous micro-structures. In this work, we present a theoretical and numerical study of graded and anisotropic porous materials, for optimal broadband and angular absorption. Through a parametric study, the effective acoustic and geometric parameters of homogenized anisotropic unit cells constitute a database in which the optimal anisotropic and graded material will be searched for. We develop an optimization technique based on the simplex method that is relying on this database. The concepts of average absorption and diffuse field absorption coefficients are introduced and used to maximize angular acoustic absorption. Numerical results present the optimized absorption of the designed anisotropic and graded porous materials for different acoustic targets. The designed materials have anisotropic and graded effective properties, which enhance its sound absorption capabilities. While the anisotropy largely enhances the diffuse field absorbing when optimized at a single frequency, graded properties appear to be crucial for optimal broadband diffuse field absorption.

Keywords: anisotropic materials; optimized absorption; diffuse field; graded properties

1. Introduction

In the context of acoustic wave propagation, porous structures are commonly employed for sound absorption [1–3]. Their efficiency has been demonstrated many times over the past decades and they have been exploited in numerous applications, like civil engineering, room acoustics and building insulation. They are particularly useful for compact designs and cheap manufacturing. Their ability to absorb sound is often characterized by the diffuse field absorption coefficient which is defined as a weighted average of the absorption for all possible angles of incidence [4–7]. What would be considered as an optimal absorber would see its absorption coefficient maximized for all frequencies, under all possible angles of incidence. In order to envisage such an absorber, the design and optimization of anisotropic and graded materials are explored in this work.

Porous materials consist in two-phase media, in which a solid phase Ω_s, here considered rigid and motionless is saturated by a fluid phase Ω_f, in our case air [8]. The open porosity is denoted by the scalar $\phi = |\Omega_f|/|\Omega_f \cup \Omega_s| \in [0,1]$, and accounts for the ratio of the fluid volume over that of the total domain. In these conditions, the porous material can be treated as, whose acoustical properties depend on the micro-structure. Interestingly, as soon as the micro-structure is known, as for example in porous materials made of periodic arrangements of unit cells, the parameters of the Johnson–Champoux–Allard–Lafarge (JCAL) model can be efficiently computed by a two-scale asymptotic method. These parameters are the open porosity, high-frequency limit of tortuosity, characteristic thermal and viscous lengths, and static thermal and viscous permeabilities. The high-frequency limit of tortuosity can be interpreted as the path of a particle, in the inertial regime of the fluid. Both thermal and viscous characteristic lengths are related to thermal and viscous skin depth, and the size of the pores. Finally, the static viscous and thermal permeabilities are linked to the flow restitivty and compressibility. In the case of non-isotropic materials, the tortuosity, the characteristic viscous length and the static viscous permeability admit different values in each direction. The definition of these coefficients and their physical interpretation is detailed in Ref. [9]. The thermal and viscous dynamic regimes, occurring inside the medium, arise from the fluid-structure interaction at the microscopic scale of the unit cell [10]. The JCAL model uses frequency asymptotics to estimate these complex and frequency-dependent effective permeabilities of the porous medium described by the six parameters. In the most general case, the unit cells can be anisotropic and thus, the equivalent fluid can display anisotropic features. In this case, the mass density of the effective medium will be denoted by the symmetric, second-order tensor $\boldsymbol{\rho}_{(e)}$ which accounts for in-plane and normal direction properties, together with the bulk modulus $B_{(e)}$. Such physical modeling has been demonstrated and validated experimentally in many instances [11,12]. The method consists in the application of a two-scale asymptotic homogenization to governing fundamental equations. The JCAL parameters are then calculated by integrating the computed fields [13–15].

In addition, the effective properties describing the wave propagation in the medium can also vary along a specific direction, specifically, together with the complex and frequency dependence, the medium displays spatially dependent features. Therefore, altering the micro-structure's geometry at different locations affects the propagation of acoustic waves. In a rigorous manner, the solution of an anisotropic, spatially dependent Helmholtz equation describes the acoustic propagation inside the porous material. In this regard, various analytical and numerical methods of acoustic wave propagation have been proposed, for multi-layered and continuously space varying materials. Wave-splitting techniques with Green's functions (WSTGF) or Peano series expansions (PS) can be applied for this purpose [16–22]. In order to reduce the computational cost, the well-known transfer matrix method (TMM) has been also used [23,24].

Recent work on optimized porous materials has been devoted to two main goals. On the one hand, maximizing sound absorption [14,22], and on the other hand recovering the JCAL parameters by developing inverse characterization methods [13]. It has been recently shown that quasi-perfect broadband absorption can be achieved at normal incidence, using graded porous layers [14]. In this work we present the optimization of anisotropic and graded materials for oblique incidences, paying special attention to the omnidirectional and broadband absorption. This relates to the property of maximal and uniform absorption for all possible angles of incidence [25]. Within the frame of this work, we use a set of unit cells which are described by their geometric parameters. In order to model the propagation in such materials, a database of JCAL parameters is established by means of finite elements method (FEM), as in Refs. [13–15]. We use a database of 100 anisotropic unit cells, which links the JCAL parameters to the geometric ones, for every unit cell. Starting from this set of homogenized anisotropic unit cells, variations of the geometry are introduced along the depth of the equivalent fluid layer, so as to achieve prescribed targets for the frequency and angular absorption. These targets are defined through the introduction of cost functions that are minimized based on the average and diffuse field absorption coefficients.

The present article is organized as follows: we first recall the general derivation of wave propagation in anisotropic and graded fluid layers in Section 2. We then discuss the interpretation of average and diffuse field absorption in Section 3. In Section 4 the data generation and the optimization procedures are described in detail. Finally, multiple numerical results are presented and discussed in Section 5 and are compared to those of isotropic and non-graded media, to demonstrate the benefits of anisotropic and graded materials.

2. Reminder on Anisotropic and Graded Porous Materials

In this section the propagation of a plane wave through an anisotropic, uni-dimensional (1D) graded equivalent fluid is recalled. More in-depth knowledge and overall derivation of the equations can be found in the Refs. [13,14,22]. We set the Cartesian coordinate system $\mathcal{R}_0 = (O, \mathbf{e}_1, \mathbf{e}_2, \mathbf{e}_3)$ with the associated spatial coordinates vector $\mathbf{x} = (x_1, x_2, x_3) \in \mathbb{R}^3$ as defined in Figure 1. The equivalent fluid domain, denoted $\Omega_e \equiv \Omega_f \cup \Omega_s$, is a slab of finite thickness L in the x_3 direction and of infinite extent in the $(O, \mathbf{x}_\perp)$ plane. The subscript $_\perp$ denotes the restriction of a vector or tensor to the $(O, \mathbf{x}_\perp)$ plane with $\mathbf{x}_\perp = \{x_1, x_2\}$. The domain Ω_e is bounded by the plane boundaries at $x_3 = 0$ and $x_3 = L$ denoted Γ_0 and Γ_L respectively. We solve for the sound field in this layer Ω_e in the linear harmonic regime using the time convention $\mathrm{e}^{-\mathrm{i}\omega t}$ where $\omega = 2\pi f$ is the angular frequency. The exterior of the domain Ω_e is denoted Ω_0 and corresponds to $x_3 > L$. It contains an homogeneous isotropic fluid, taken to be air in this case and considered inviscid. The density of air is $\rho_{(0)} = 1.213$ kg·m^{-3} and its bulk modulus $B_{(0)} = \gamma P_0$ with $\gamma = 1.4$ the ratio of specific heat and $P_0 = 101{,}325$ Pa the atmospheric pressure.

The pressure p and velocity $\mathbf{v}$ induced by the acoustic field are governed by the following linear equations for mass conservation and momentum conservation respectively,

$$\mathrm{i}\omega \boldsymbol{\rho}_{(j)}(x_3, \omega)\mathbf{v}(\mathbf{x}, \omega) = \nabla p(\mathbf{x}, \omega)\,, \tag{1a}$$

$$\mathrm{i}\omega p(\mathbf{x}, \omega) = B_{(j)}(x_3, \omega)\nabla \cdot \mathbf{v}(\mathbf{x}, \omega)\,, \tag{1b}$$

where the subscript $j = \{0, e\}$ designates the domains Ω_0 and Ω_e respectively. The effective bulk modulus and mass density tensor of the anisotropic 1D graded equivalent fluid (along the $\mathbf{e}_3$ direction) are denoted $B_{(e)}(x_3, \omega)$ and $\boldsymbol{\rho}_{(e)}(x_3, \omega)$. In the particular case of a transverse isotropic material, the density tensor can be written $\boldsymbol{\rho}_{(e)} = \mathrm{diag}(\rho_\perp, \rho_\perp, \rho_3)$ in its principal directions. Note that these quantities are complex-valued, frequency-dependent and can vary along the $\mathbf{e}_3$ direction. Moreover, while the bulk modulus of the anisotropic medium is scalar, the mass density is a second-order tensor accounting for the anisotropy. To summarize, a non-graded anisotropic material is modeled by the physical quantities $\boldsymbol{\rho}_{(e)}(\omega)$ and $B_{(e)}(\omega)$, whereas for graded anisotropic materials, these quantities $\boldsymbol{\rho}_{(e)}(x_3, \omega)$ and $B_{(e)}(x_3, \omega)$ are spatially dependent.

In the air region Ω_0 ($x_3 \geq L$) we define an incident plane wave with unit amplitude:

$$p^{inc}(\mathbf{x}, \omega) = \mathrm{e}^{\mathrm{i}\mathbf{k}_\perp \cdot \mathbf{x}_\perp - \mathrm{i}k_3(x_3 - L)}\,, \tag{2}$$

where the components of the wave-vector $\mathbf{k}^{inc} = \{k_1, k_2, k_3\}^T$ (with ^{T} the non-conjugate transpose) are given by

$$\begin{cases} k_1 = -k_0 \cos(\theta)\cos(\psi), \\ k_2 = -k_0 \cos(\theta)\sin(\psi), \\ k_3 = k_0 \sin(\theta), \end{cases} \tag{3}$$

with ψ and θ the polar and elevation angles, respectively. The free-field acoustic wave-number is $k_0 = |\mathbf{k}^{inc}| = \omega/c_0$, and the sound speed in free air is defined by $c_0^2 = B_{(0)}/\rho_{(0)}$.

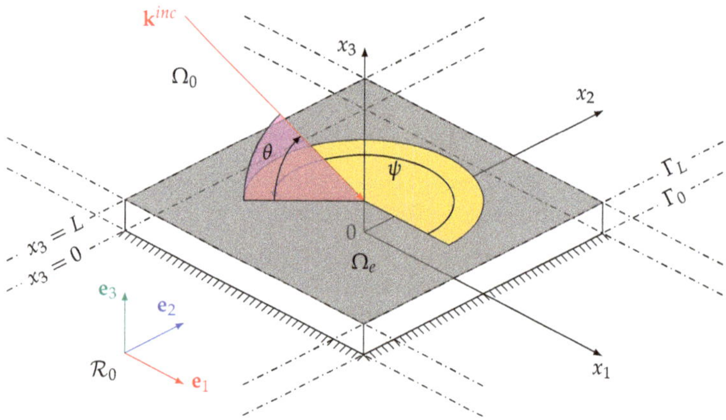

Figure 1. (Color online) Schematic of the propagation problem. The incident wave-vector $\mathbf{k}^{inc}$ has azimuth and elevation angles ψ (yellow) and θ (light purple) respectively.

The presence of the anisotropic layer Ω_e gives rise to a reflected wave p^r in Ω_0. It is written

$$p^r(\mathbf{x}, \omega) = R e^{\mathrm{i}\mathbf{k}_\perp \cdot \mathbf{x}_\perp + \mathrm{i}k_3(x_3 - L)}, \tag{4}$$

where $R(\omega, \theta, \psi)$ is the specular reflection coefficient and $\mathbf{k}_\perp^{inc} = \{k_1, k_2\}^T$. The derivation of the governing equations Equation (1) has recently been used for retrieval techniques and applied to fully-anisotropic porous materials [13]. They can be written using a state-vector formulation,

$$\frac{\mathrm{d}\mathbf{W}}{\mathrm{d}x_3} = \mathbf{A}(x_3)\mathbf{W}, \tag{5}$$

where we have introduced the state vector $\mathbf{W} = \{p, v_3\}^T$, and the matrix

$$\mathbf{A}(x_3) = \begin{bmatrix} 0 & \mathrm{i}\omega\rho_3 \\ \mathrm{i}\omega B_{eq}^{-1} & 0 \end{bmatrix}, \tag{6}$$

which admits no diagonal terms when the principal directions of the fluid are aligned to those of the coordinate system. The propagation problem is reduced to a system of equations with respect to x_3. The term B_{eq} is the equivalent bulk modulus and accounts for transverse $(O, \mathbf{x}_\perp)$ effective density and oblique incidence. In the case of anisotropic materials, with mass density tensor $\boldsymbol{\rho}_{(e)} = \mathbf{H}^{-1}, \mathbf{H} \in \mathbb{C}^{3\times3}$, bulk modulus $B_{(e)}$ and wave-vector $\mathbf{k}$, we have,

$$B_{eq}^{-1} = B_{(e)}^{-1} - [\mathbf{k}_\perp \cdot (\mathbf{H}_\perp \cdot \mathbf{k}_\perp)] / \omega^2. \tag{7}$$

At the interface Γ_L between the anisotropic layer and the surrounding fluid, the continuity of pressure and normal velocity is imposed as boundary conditions. At the interface Γ_0, a rigid backing is considered in this work, and thus zero normal particle velocity is imposed by $\mathbf{v} \cdot \mathbf{n}|_{\Gamma_0} = 0$. As a consequence, the state-vector at both interfaces reads,

$$\mathbf{W}_L = \left\{ \begin{array}{c} 1 + R \\ (R-1)/Z_0^\theta \end{array} \right\} \quad \text{and} \quad \mathbf{W}_0 = \left\{ \begin{array}{c} p(0) \\ 0 \end{array} \right\}, \tag{8}$$

with $Z_0^\theta = Z_0 / \sin(\theta)$ the apparent impedance of the air in domain Ω_0 with respect to the unit outward normal vector $\mathbf{n} = \mathbf{e}_3$ at interface Γ_L. The system of equations in Equation (5) can be solved by various techniques, being the TMM, PS and WSTGF as in Refs. [16–24]. This way, we retrieve the reflection coefficient of graded and anisotropic porous materials with a rigid backing. The absorption coefficient is defined with respect to the angular frequency, and angles of incidence as

$$\alpha(\omega,\theta,\psi) = 1 - |R(\omega,\theta,\psi)|^2 \,. \tag{9}$$

To sum up, from a given geometry of the unit cell, one can compute the JCAL parameters as in Refs. [13–15] using FEM. From the JCAL parameters, we obtain the mass density tensor $\boldsymbol{\rho}_{(e)}(\omega)$ and bulk modulus $B_{(e)}(\omega)$ of the homogenized medium [9]. As these effective properties depend on the geometry of the porous material, we can introduce variations of the geometric properties along the direction x_3 by using $\boldsymbol{\rho}_{(e)}(x_3,\omega)$ and $B_{(e)}(x_3,\omega)$. Then, the reflection coefficient is provided by solving Equation (5), using the aforementioned numerical techniques.

3. Average and Diffuse Field Absorptions

In the Cartesian coordinate system $\mathcal{R}_0 = (O, \mathbf{e}_1, \mathbf{e}_2, \mathbf{e}_3)$, the incident wave is represented by its wave-vector $\mathbf{k}^{inc} = \{k_1, k_2, k_3\}^T$. The diffuse sound field is composed of a continuum of plane waves evenly distributed over the elevation angle θ and azimuth angle ψ. These plane waves have the same acoustic intensity and are uncorrelated [4,6,7]. To define an absorption coefficient for this diffuse field, the absorption coefficient $\alpha(\omega,\theta,\psi)$ defined above, in Equation (9) for an individual plane wave has to be averaged over all possible incidence angles. In this work, two distinct cases will be considered regarding this averaging of the acoustic absorption coefficient. The first average is based on the standard definition of the diffuse field absorption coefficient, which reads as follows [5],

$$\alpha_{dif}(\omega) := \frac{\displaystyle\int_0^{\pi/2}\int_0^{2\pi} \alpha(\omega,\theta,\psi)\sin(\theta)\cos(\theta)\mathrm{d}\psi\mathrm{d}\theta}{\displaystyle\int_0^{\pi/2}\int_0^{2\pi} \sin(\theta)\cos(\theta)\mathrm{d}\psi\mathrm{d}\theta}\,, \tag{10}$$

where $\sin(\theta)\mathrm{d}\theta\mathrm{d}\psi$ is due to the change in area of the surface integration element on the sphere. In addition, the component of the acoustic intensity vector pointing into the surface is proportional to $\cos(\theta)$. It follows that the normal and grazing angles of incidence are do not contribute in this definition, as illustrated in Figure 2b by the blue dashed line. The integral formalism in Equation (10) can be approached by a discrete sum. As described earlier in Section 2, we use transverse isotropic materials. Hence the polar angle ψ has no influence on the absorption, we now have,

$$\alpha_{dif}(f) \approx \left(\sum_{i=1}^{N_\theta} \alpha(f,\theta_i)\sin(\theta_i)\cos(\theta_i)\right)\left(\sum_{i=1}^{N_\theta}\sin(\theta_i)\cos(\theta_i)\right)^{-1}, \tag{11}$$

where N_θ is the number of elevation angles. The second average is based on the arithmetic average of the absorption coefficient, illustrated in Figure 2c:

$$\bar{\alpha}(f) \approx \frac{1}{N_\theta}\sum_{i=1}^{N_\theta}\alpha(f,\theta_i)\,. \tag{12}$$

We note that when compared to the diffuse field average Equation (10), the angular weighting related to the variations of the solid angle and normal acoustic intensity are not accounted for in Equation (12).

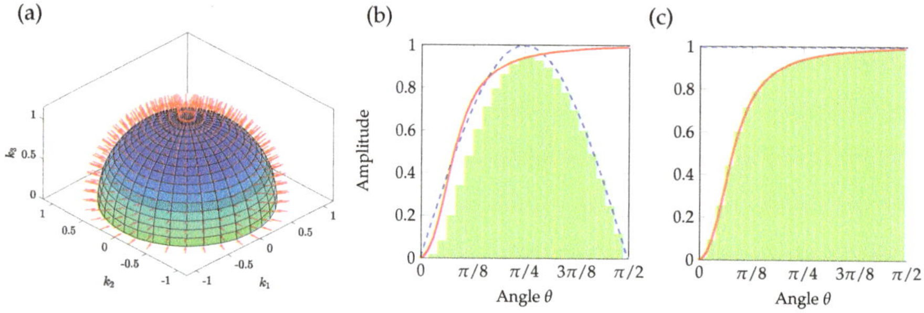

Figure 2. (Color online) Weighting process for average and diffuse field absorption coefficient. (**a**) Discrete incidences on the unit sphere and elementary surfaces. The subfigures (**b**,**c**) represent the integration process defined in Equations (11) and (12) respectively. The solid red line is an example of an angular-dependent absorption coefficient, the weighting function is shown in a dashed blue and the light green bars illustrate the discrete integration.

4. Optimization Procedures

The optimization routines provide the anisotropic and graded properties of the porous layer for a given acoustic target. In particular, for a fixed layer thickness L, the macro-modulated effective density $\boldsymbol{\rho}_{(e)}$ and bulk modulus $B_{(e)}$ are obtained for the given acoustic absorption properties.

4.1. Interpolated Database of Unit Cells

The proposed unit cell is made of an ellipsoid carved out from a rectangular cuboid, as illustrated in Figure 3i. This unit cell has already been used for retrieval methods in Ref. [13]. To define this geometry, several parameters are introduced, namely the open porosity ϕ, characteristic length ℓ_c and stretching in each in-plane direction χ_1 and χ_2. For simplicity, the two stretching parameters are set to be equal: $\chi_1 = \chi_2 = \chi$. A set of 100 unit cells with $\phi \in [0.56, 0.89]$, and $\chi \in [1, 10]$ is first used to build up a database of the JCAL parameters, representing 100 anisotropic porous materials made of the periodic repetition of the unit cell, with values (ϕ, χ_1, χ_2) from the previous intervals. The interval for porosity is driven by the topology of the unit cell, whereas the stretch is chosen to span one order of magnitude. These properties are obtained through a homogeneization method relying on finite elements methods (see [13,14] for more details). The graded anisotropic materials used in this work for the optimization of the acoustic absorption properties can be composed of layers of different materials, the values of porosity and stretching of each layer might correspond to a periodic porous material made of one of the unit cells in the given database.

In the case in which the couple (ϕ, χ) does not correspond to one of the 100 unit cells, the effective properties are obtained by interpolating between the data points existing in the database of JCAL parameters. This is done using piecewise cubic Hermite interpolating polynomials (PCHIP) [26]. The cubic interpolant of the transport parameters with the geometric parameters is monotonic. The use of such an interpolation method in this context is therefore robust. Figure 3 shows the dependency of the JCAL parameters on the stretching factor χ and open porosity ϕ. This interpolation spans the whole database and gives a set of JCAL parameters for the required unit cell, allowing us to obtain the corresponding mass density tensor and bulk modulus of the anisotropic porous material. We denote the vector of geometric parameters by $\mathcal{W}_G = \{\ell_c, \phi, \chi\}^T$ with which the unit cell is described. The corresponding JCAL parameters are stored in the vector of porous properties $\mathcal{W}_J = \{\phi, \Theta^0, \Lambda', \Lambda_\perp, \Lambda_3, \tau_{33}^\infty, \tau_\perp^\infty, K_\perp^0, K_{33}^0\}^T$.

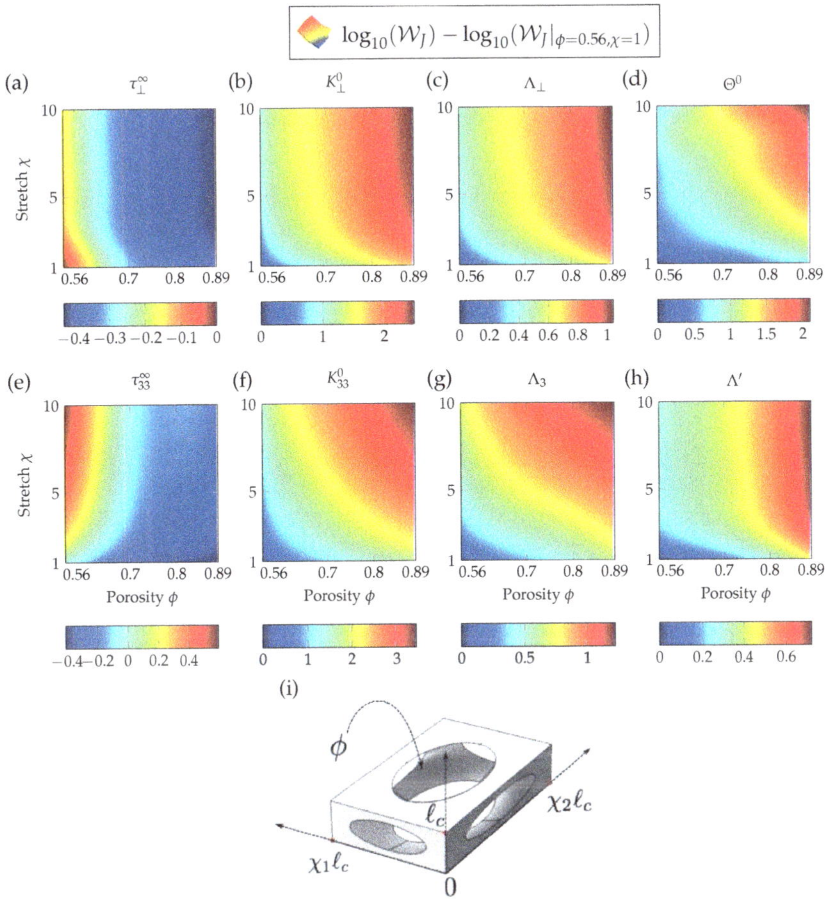

Figure 3. (Color online) Interpolated database of normalized JCAL parameters, with respect to the geometric properties ϕ and χ. The values are normalized to those of the isotropic cell with lowest porosity (corresponding to $\phi = 0.56$, $\chi_1 = \chi_2 = \chi = 1$) and are displayed in a logarithmic scale. The values in the transverse direction of the tortuosity, static viscous permeability and characteristic viscous length are shown respectively in (**a**–**c**); the normal components are displayed in (**e**–**g**). Finally, the static thermal permeability and characteristic thermal length are given in (**d**,**h**) respectively. The unit cell is displayed along with its geometric assets in (**i**).

The transport parameters shown in Figure 3, are computed for a unit characteristic length $\ell_c = 1$ m. The parameters are then scaled independently according to their dimension, in meters for Λ', $\Lambda_\perp$ and Λ_3, in square meters for Θ^0, $K_\perp^0$ and K_{33}^0. They respectively denote the thermal characteristic length, the viscous characteristic length in transverse and normal directions, the static thermal permeability, and finally the static viscous Darcy [27] permeability in transverse and normal directions. However, the open porosity ϕ and high-frequency limit of tortuosity τ_{jj}^∞ are dimensionless and independent of the characteristic length.

4.2. Macro-Modulation of Effective Properties

The fluid layer Ω_e is split in $N-1$ intervals, bounded by N linearly placed nodes along $\mathbf{e}_3$. At Γ_0 there is the first node $x_3^{(1)} = 0$, and at Γ_L we place the N-th node at $x_3^{(N)} = L$. Between them, we place the other nodes at $x_3^{(i)} = (i-1)L/(N-1), \forall i \in [\![1, N-1]\!]$. The interpolation of the graded properties based on this discretization is done using PCHIP [26]. This ensures the values at the nodes $\mathcal{W}_G\left(x_3^{(i)}\right)$ and the interpolated function is of class $\mathcal{C}^1$, meaning that its derivative remains continuous. On the interval $\left[x_3^{(i)}, x_3^{(i+1)}\right]$ between two nodes, each geometric property g (components of $\mathcal{W}_G$) reads,

$$\forall x_3 \in \left[x_3^{(i)}, x_3^{(i+1)}\right], x_3^{(i)} < x_3^{(i+1)}, \ \mathcal{W}_g(x_3) := \sum_{n=0}^{3} a_{n,g}\left(x_3 - x_3^{(i)}\right)^n, \tag{13}$$

with $a_{n,g}$ being the weights of the polynomial interpolation function. It is important for the reader to note that two different interpolations are taking place in the procedure. The first one described in the previous subsection, to interpolate the database of unit-cells, and a second one applied between spatial nodes to interpolate the graded properties of the medium Ω_e.

4.3. Optimization of Geometric Properties

In order to optimize the geometric parameters at the nodes $\mathcal{W}_G\left(x_3^{(i)}\right)$, the Nelder–Mead (or downhill simplex) method was used. It is an iterative scheme that minimizes a non-linear cost function, based on a geometric approach [28,29]. From an initial set of geometric nodes $\mathcal{W}_G\left(x_3^{(i)}\right)$, the method takes a series of steps that tend to minimize the cost function. At each iteration of the method, the set of parameters from the previous iteration is transformed within the space of parameters by means of a simplex. The scheme stops when the convergence is reached (with respect to absolute tolerances) or when the maximum iteration number is attained. The stopping condition is given when the iteration number exceeds $n = 2 \times 10^3$, or when both variations of the cost function $\partial_n J$ and of the solution vector $\partial_n \mathcal{W}_G$ do not differ by 1×10^{-6} between two iterations. The optimization is performed using the function fminsearch within MATLAB®2016b. This optimization procedure is bounded so that, as the method scans the space of parameters, the simplex can only exist within the parameter range of the database. The characteristic length was chosen between $\ell_c = 0.3$ mm and $\ell_c = 1.5$ mm, and the stretching factor between $\chi = 1$ and $\chi = 10$. In addition, the open porosity was bounded due to the cell topology, its upper and lower bounds can be derived analytically. The components of $\mathcal{W}_G$ must therefore satisfy the following constraints,

$$\begin{cases} \phi \in [0.56, 0.89], \\ \chi \in [1, 10], \\ \ell_c \in [300, 1500] \times 10^{-6} \text{ m}. \end{cases} \tag{14}$$

The initial guess used to start the optimization procedure is formed of random geometric parameters, according to a uniform distribution within the proposed bounds. The routine also has to be run multiple times with different initial guesses, to avoid local minima. On the one hand, for the optimization of the non-graded porous material case, where the vector of geometric parameters $\mathcal{W}_G$ is constant along $\mathbf{e}_3$, no interpolation is needed and the acoustic propagation problem can be solved exactly using a single layer by the TMM. On the other hand, when the properties are graded along $\mathbf{e}_3$, it is performed using an interpolation over $N = 5$ nodes, and solved with 40 layers using again the TMM. For each layer, the interpolated database connects $\mathcal{W}_G(x_3)$ and $\mathcal{W}_J(x_3)$, as to reach $\rho_{(e)}(x_3, \omega)$ and $B_{(e)}(x_3, \omega)$. This way, the acoustic features of the graded and non-graded materials are driven by the geometric properties of the porous structures.

Two distinct cost functions are introduced. The first one is based on the diffuse field absorption,

$$J_{dif} := \frac{1}{N_f} \sum_{i=1}^{N_f} \left(\alpha_{dif}(f_i) - 1\right)^2, \tag{15}$$

while the second one is based on the arithmetic average of the absorption,

$$\bar{J} := \frac{1}{N_f} \sum_{i=1}^{N_f} \left(\bar{\alpha}(f_i) - 1\right)^2 , \tag{16}$$

with f_{min} and f_{max} the lower and upper frequency bounds, respectively. In the frequency domain, the minimization operates on N_f linearly spaced values so that $\Delta f = (f_{max} - f_{min})/(N_f - 1)$. Both the diffuse field absorption Equation (11) and the average absorption Equation (12) are studied, using a linear discretization over the elevation angle $\theta \in [\pi/30, \pi/2]$ with $N_\theta = 30$ points. It gives $\Delta\theta = (\pi/2 - \pi/30)/30 \approx 0.06$ rad as the angular elevation step over all possible incidences.

Considering each cost function, the algorithm will maximize the absorption within targets of frequency and elevation angle. The vector of geometric parameters $\mathcal{W}_G$ is optimized to create graded and non-graded materials. The database that links the geometric parameters $\mathcal{W}_G$ with the JCAL parameters $\mathcal{W}_J$ is used to obtain Equation (5). This equation is finally solved using TMM as it is usually faster, and provides the absorption coefficient for the graded and non-graded materials.

5. Results

Results of the optimization procedure are shown for different acoustic targets, in order to highlight the benefits of anisotropic and graded effective properties with respect to the isotropic and non-graded ones. We first display the results in the case of a single frequency, when the porous layer operates in its sub–wavelength regime, that is, when the wavelength of the incident wave was larger than the thickness of the porous material ($\lambda = c_0/f > L$). In this work, we consider the length $L = 25$ mm. Then, the optimization was applied to broadband acoustic absorption, on the frequency range between 1 kHz and 5 kHz, in other words, $\lambda/L = 13.6$ and $\lambda/L = 2.7$ respectively.

5.1. Sub–Wavelength Acoustic Absorption

We consider a single frequency $f = 1000$ Hz at which the monochromatic plane wave impinges the porous layer either with anisotropic or isotropic properties. The cost functions described earlier in Equations (15) and (16) do not need the sum over frequency for this optimization.

Figure 4a,b,d,e show the average absorption coefficients and the diffuse field absorption coefficients for anisotropic (isotropic) materials with graded and non-graded properties optimized using J_{dif} and $\bar{J}$, respectively. In all cases, we can see that both the average absorption and the diffuse field absorption (see Figure 4a,b,d,e) present highest values for the graded porous materials at the target frequency.

The difference between the absorption of graded and non-graded materials is more pronounced for isotropic than for anisotropic materials. This result is important to highlight the relevance of the additional degree of freedom introduced by the spatially dependent properties. On the other hand, it is important to compare the values of absorption between the anisotropic and the isotropic materials. In that regard, we can clearly see the improvements of the absorption by the anisotropic materials showing the importance of the added degree of freedom by this feature. For both $\bar{J}$ and J_{dif}, the non-graded medium provides excellent results in terms of absorption. As seen on Figure 4a,b, the absorption peak at $f = 1000$ Hz reaches $\bar{\alpha} = 0.94$ and $\alpha_{dif} = 0.97$. Figure 4c,f shows the angular profile of absorption at $f = 1000$ Hz for anisotropic (isotropic) materials with graded and non-graded properties optimized using J_{dif} and $\bar{J}$. These results show one important feature related to the impact of the cost function on the angular absorption profiles. While optimizing the average absorption does not promote any specific angle (the cost function $\bar{J}$ accounts for all incidences equally), optimizing the diffuse field absorption coefficient does not account for the grazing and normal angles of incidence. Looking at the average absorption, we see that it is high at normal incidence, and creates a plateau until it finally decreases after $\theta \approx \pi/8$ rad as shown in Figure 4c. However, from the optimization for the diffuse field absorption, we can see the smaller values at normal and grazing incidences as the other incident angles present higher weight in the optimization. By comparing Figure 4c,f we notice the improvement of the results by using anisotropic materials.

In the case of anisotropic materials, where the diffuse field and average absorption are strong, grading the properties of the layer along its depth does not help significantly. Looking at the JCAL parameters that give such absorption in Figures 5c and 6c, the medium displays significant anisotropic features. As seen by the high-frequency limit of tortuosity and static viscous permeability, which are one order of magnitude higher in the normal direction $\mathbf{e}_3$ than in the in-plane ones $\mathbf{e}_\perp$. Moreover, the snapshots of the unit cells in Figures 5d and 6d, display the shape in the optimized unit cells with large stretch χ, resulting in higher normal tortuosity.

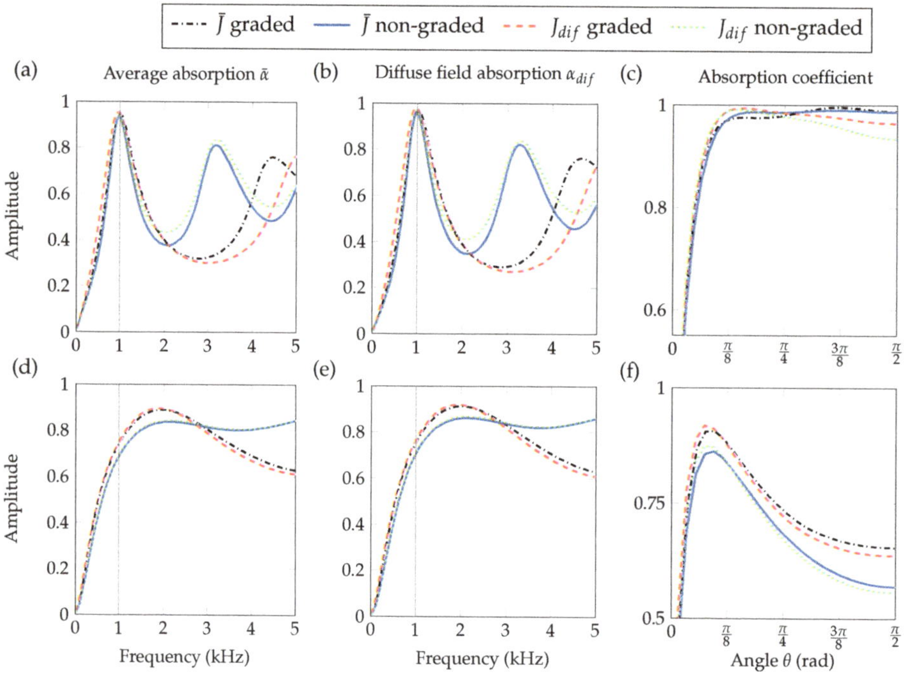

Figure 4. (Color online) Comparison of different optimization results for average and diffuse field absorption coefficients. The angular dependence of the absorption coefficient is given at the frequency $f = 1000$ Hz, with respect to θ. Results with anisotropic cells are given in (**a–c**), while the restriction to isotropic cells is shown in (**d–f**). In solid red and dashed blue are the results for $\bar{J}$ with $N = 1$ and $N = 5$, respectively, and for J_{dif} with $N = 1$ in dashdotted green and $N = 5$ in dotted black.

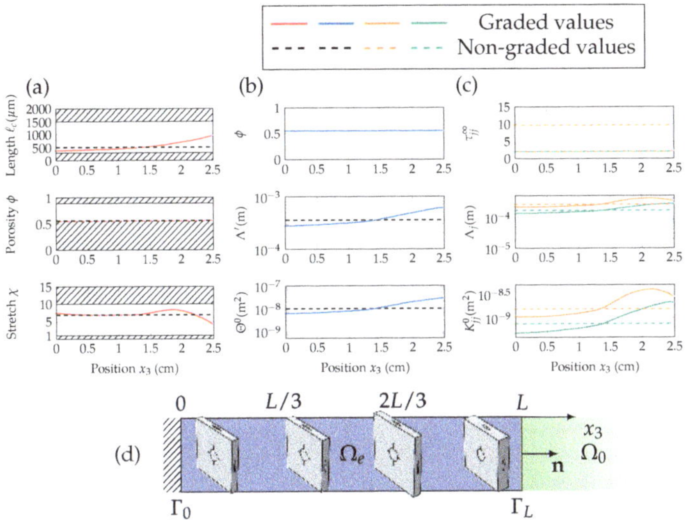

Figure 5. Components of the vector of geometric parameters $\mathcal{W}_g$ (characteristic length ℓ_c, porosity ϕ and stretch χ) are shown in (**a**), as a function of the position x_3. Scalar Johnson–Champoux–Allard–Lafarge (JCAL) parameters (ϕ, Θ^0, Λ') are given in (**b**) and the direction-dependent ones ($\tau_{\perp}^{\infty}, \tau_{33}^{\infty}, K_{\perp}^{0}, K_{33}^{0}, \Lambda_{\perp}, \Lambda_3$) are in (**c**), with respect to x_3. The graded values are in solid lines, and the non-graded ones in dashed lines; in orange for the normal direction, and in green for the transverse direction. The sketch in (**d**) displays snapshots of the unit cells at intervals of $L/3$ between interfaces Γ_0 and Γ_L.

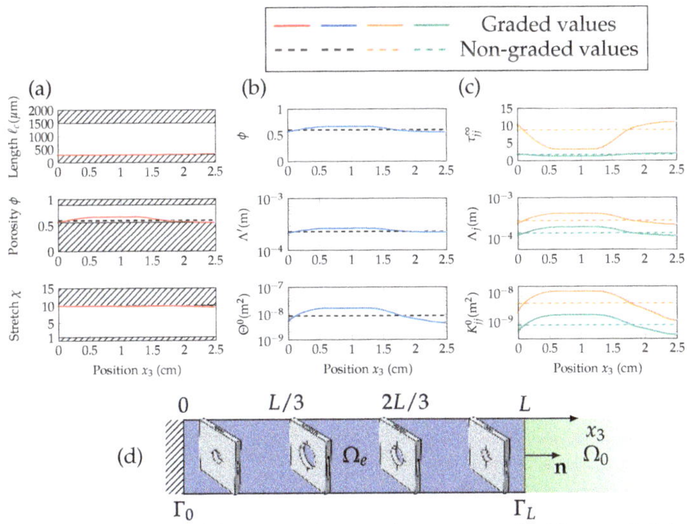

Figure 6. Components of the vector of geometric parameters $\mathcal{W}_g$ (characteristic length ℓ_c, porosity ϕ and stretch χ) are shown in (**a**), as a function of the position x_3. Scalar JCAL parameters (ϕ, Θ^0, Λ') are given in (**b**) and the direction-dependent ones ($\tau_{\perp}^{\infty}, \tau_{33}^{\infty}, K_{\perp}^{0}, K_{33}^{0}, \Lambda_{\perp}, \Lambda_3$) are in (**c**), with respect to x_3. The graded values are in solid lines, and the non-graded ones in dashed lines; in orange for the normal direction, and in green for the transverse direction. The sketch in (**d**) displays snapshots of the unit cells at intervals of $L/3$ between interfaces Γ_0 and Γ_L.

For the sake of comparison, the optimization is also performed with a restriction on the database to isotropic cells, only with $\chi = 1$. This restriction implies that the medium cannot take contrasted properties in its principal directions. In this case, using graded properties becomes much more important to achieve a strong absorption, and we reach $\alpha_{dif} = 0.76$ and $\bar{\alpha} = 0.74$. In Figure 4d,e, the results for an isotropic porous material show the benefit of the geometric gradient $\mathcal{W}_G(x_3)$, when the medium is constrained to be isotropic.

5.2. Broadband Acoustic Absorption

The frequency vector is given by $N_f = 15$ linearly spaced values between $f_{min} = 1000$ Hz and $f_{max} = 5000$ Hz. This frequency range ensures the large wavelength condition as we have $\lambda > 2L$ at $f = f_{max}$. This is accounted for in the minimization of the cost functions Equations (15) and (16), by the sum over the frequencies. No weighting is considered in frequency, meaning that all the frequencies have the same importance. The goal of the optimization routine is to minimize the reflection over the whole angular and frequency plane (θ, f), discretized in $N_f \times N_\theta = 450$ points.

Regarding the geometric and JCAL profiles shown in Figures 7a and 8a, we see that the optimized non-graded material is isotropic: $\chi = 1$. In this case, over the frequency range of interest, the high-frequency limit of normal tortuosity τ_{33}^{∞} does not need to reach important values, for the absorption to reach $\alpha > 0.9$. Figures 7c and 8c show the successive troughs and peaks of the normal tortuosity, when the medium is graded along x_3. This trend is observed when minimizing for both cost functions, and is intuitively represented on Figures 7d and 8d, as smaller pores and higher transverse stretches tend to increase τ_{33}^{∞}.

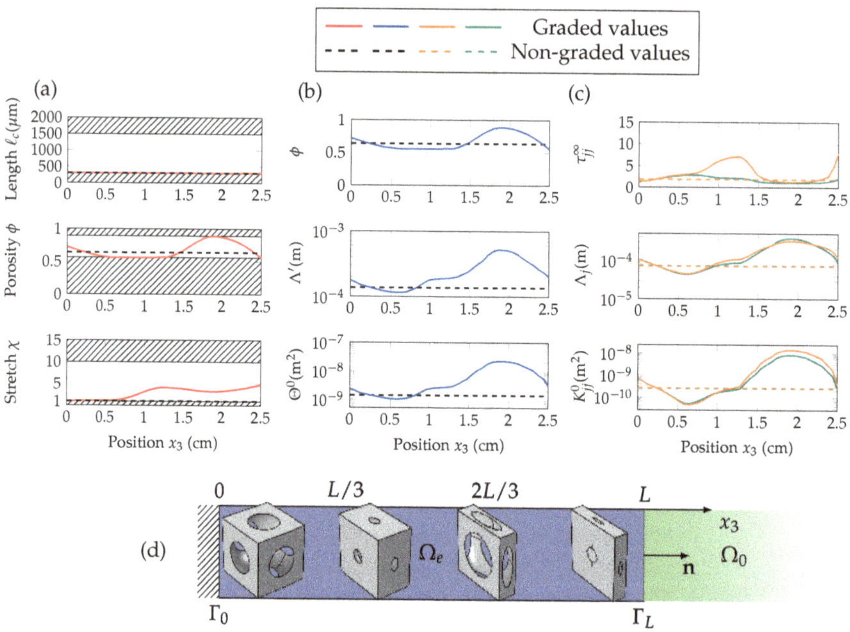

Figure 7. (Color online) Components of the vector of geometric parameters $\mathcal{W}_g$ (characteristic length ℓ_c, porosity ϕ and stretch χ) are shown in (**a**), as a function of the position x_3. Scalar JCAL parameters $(\phi, \Theta^0, \Lambda')$ are given in (**b**) and direction-dependent ones $(\tau_{\perp}^{\infty}, \tau_{33}^{\infty}, K_{\perp}^0, K_{33}^0, \Lambda_{\perp}, \Lambda_3)$ are in (**c**), with respect to x_3. The graded values are in solid lines, and the non-graded ones in dashed lines; in orange for the normal direction, and in green for the transverse direction. The sketch in (**d**) displays snapshots of the unit cells at intervals of $L/3$ between interfaces Γ_0 and Γ_L.

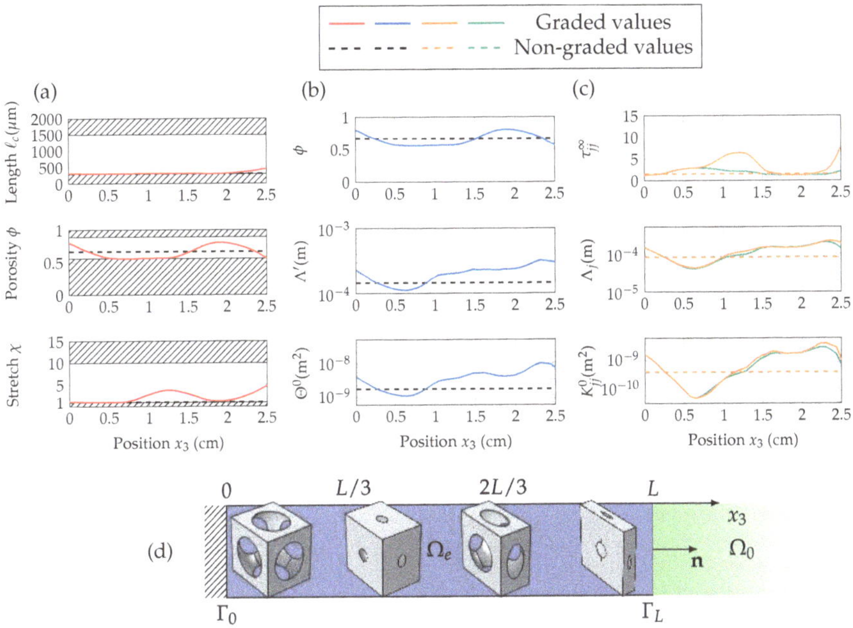

Figure 8. (Color online) Components of the vector of geometric parameters $\mathcal{W}_g$ (characteristic length ℓ_c, porosity ϕ and stretch χ) are shown in (**a**), as a function of the position x_3. Scalar JCAL parameters (ϕ, Θ^0, Λ') are given in (**b**) and direction-dependent ones ($\tau_\perp^\infty, \tau_{33}^\infty, K_\perp^0, K_{33}^0, \Lambda_\perp, \Lambda_3$) are in (**c**), with respect to x_3. The graded values are in solid lines, and the non-graded ones in dashed lines; in orange for the normal direction, and in green for the transverse direction. The sketch in (**d**) displays snapshots of the unit cells at intervals of $L/3$ between interfaces Γ_0 and Γ_L.

Figure 9a,b shows the average absorption in the plane (θ, f), for the materials made of the optimal profiles with non-graded and a graded properties in the target frequency range minimizing $\bar{J}$. In the same manner, Figure 9d,e shows the diffuse field absorption in the plane (θ, f) for the materials made of the optimal profiles of a non-graded and a graded material in the target range of frequencies minimizing J_{dif} respectively. The optimized non-graded medium provides a large absorption area in the (θ, f) plane, showing a single peak of absorption in the $[f_{min}, f_{max}]$ interval (see Figure 9a,d). However, in the case of the graded material we can clearly observe two peaks of absorption in Figure 9b,e.

The optimized curves of average and diffuse field absorption for a graded material, Figure 9c,f, also clearly show this second peak of absorption for the graded materials in the $[f_{min}, f_{max}]$ interval that is absent in the case of non-graded materials. The maximum of absorption provided by the non-graded medium is at $f = (2750 \pm 25)$ Hz with values $\bar{\alpha} = 0.91$ and $\alpha_{dif} = 0.94$. In the case of graded materials optimized for the diffuse field, the absorption coefficient reaches $\alpha_{dif} = 0.93$ at the first peak at $f = (1900 \pm 25)$ Hz and $\alpha_{dif} = 0.96$ at the second peak $f = (4150 \pm 25)$ Hz, while when optimizing the average absorption, the peaks appear at $\bar{\alpha} = 0.92$ with $f = (1850 \pm 25)$ Hz and $\bar{\alpha} = 0.96$ at $f = (4100 \pm 25)$ Hz.

This clearly shows the relevance of the spatial profiles of the material that can be used to manipulate the vector of geometric properties $\mathcal{W}_G(x_3)$ introducing a direct effect on the JCAL parameters $\mathcal{W}_J(x_3)$ and influencing directly the absorption features of the system. Grading the effective properties, namely the tensor of mass density $\boldsymbol{\rho}_{(e)}(x_3)$ and bulk modulus $B_{(e)}(x_3)$, offers more control on the absorption. However, by comparing Figure 9a,d or Figure 9b,e, we can see that for broadband and angular optimization the absorption coefficient displays only small variations depending on the cost function being minimized. Finally, it is worth noting that the peak of the non-graded materials or the double peak for the graded material are still present, with the average of the diffuse absorption coefficients calculated in Figure 9c,f.

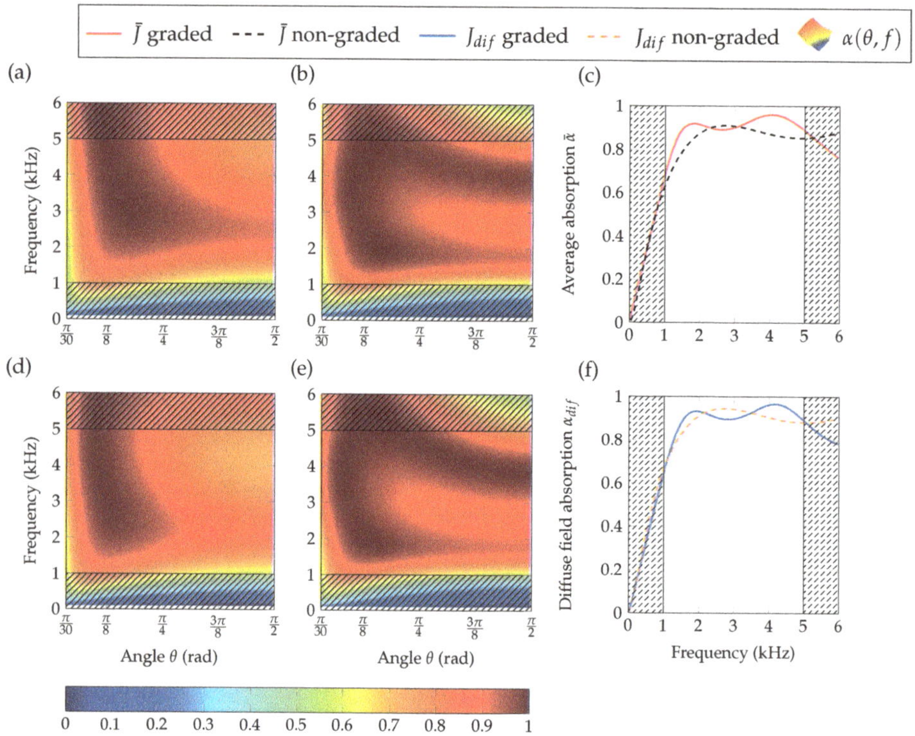

Figure 9. (Color online) Absorption maps with respect to elevation angle and frequency for broadband optimization. The results for optimized average absorption, both non-graded and graded, are shown respectively in (**a**,**b**). The maps (**d**,**e**) display the absorption map for optimized diffuse field absorption coefficient, respectively for non-graded and graded medium. Average absorption is given in (**c**), for graded and non-graded cases in solid red and dashed black respectively. The diffuse field absorption coefficient is shown in (**f**), for graded and non-graded cases in solid blue and dashed orange respectively.

At this stage, it is worth analyzing the presence of the second resonance peak in the anisotropic graded material. In order to do so, we represent the reflection coefficient in the complex frequency plane. This methodology has been widely used in the past to gain more insight into the absorption properties of acoustic materials [30,31]. In the absence of losses, this representation presents a unitary reflection in the real frequency axis, due to the conservation of energy, its resonance implies a pair of zero and poles, at complex conjugated frequencies due to the temporal invariance of the system. When losses are introduced in the system, the zeros of the reflection coefficient approach the real axis, fulfilling the impedance match condition (or the critical coupling condition) when the zero is located on the real frequency axis, producing perfect absorption. This technique has been previously described in [14] for the optimization of graded layers at normal incidence, or in [32,33] to explain how to manipulate the zeros of the reflection coefficient in order to achieve perfect absorption.

We define the complex-valued angular frequency by $\tilde{\omega} = 2\pi(f_R + \mathrm{i} f_I)$, with, $f_R \in [0, f_{max}]$, $f_I \in [-f_{max}, f_{max}]$. The differential system of equations in Equation (6) to be solved now reads,

$$\frac{\mathrm{d}}{\mathrm{d}x_3}\left\{\begin{array}{c} p \\ v_3 \end{array}\right\} - \mathrm{i}\tilde{\omega}\begin{bmatrix} 0 & \tilde{\rho}_3(\tilde{\omega}) \\ \tilde{B}_{eq}^{-1}(\tilde{\omega}) & 0 \end{bmatrix}\left\{\begin{array}{c} p \\ v_3 \end{array}\right\} = 0\,, \tag{17}$$

where $\tilde{B}_{eq}$ and $\tilde{\rho}_3$, are respectively the equivalent bulk modulus and equivalent mass density in the normal direction, at complex angular frequency $\tilde{\omega}$. The reflection coefficient $\tilde{R}(\tilde{\omega}) \in \mathbb{C}$ is computed and represented in the complex frequency plane. The perfect absorption $\alpha = 1$ is attained when the zero of complex reflection coefficient is exactly located on the real frequency axis f_R so $f_I = 0$. Therefore, we track the trajectory of the zeros for the non-graded and graded configurations, in terms of the angle of incidence θ. The angle at which the zero will be on the real frequency axis will give the angle of incidence at which the system produces perfect absorption. Figure 10a,c show the complex frequency map of $\tilde{R}$, for the non-graded optimized system at the incidence angle θ, for which the zero is on the real frequency axis, that is, at which perfect absorption is obtained by using $\bar{J}$ and J_{dif}, respectively. The trajectory of the zero crosses twice the real frequency axis, showing that two configurations present perfect absorption, but only one stands in the target range of frequencies. Figure 10b,d show the complex frequency map of $\tilde{R}$ for the graded optimized system at the incidence angle at which the first zero is on the real frequency axis, that is, the perfect absorption of the first peak by using $\bar{J}$ and J_{dif} respectively. Here, we can observe two zeros corresponding to the previous two peaks of absorption for the graded case. The trajectories of the zeros cross twice the real frequency axis, showing that there are four configurations with perfect absorption but only two are in the target range of frequencies. There is a general tendency of the zeros independently of the type of material used. In the frequency range of interest, we observe the zeros slowly drifting to higher frequencies as the angle of incidence decreases, visible on Figure 9a,b,d,e.

In contrast to the non-graded materials, the graded ones contribute to lowering the second reflection zero in the frequency range of interest, increasing the absorption across the entire range of frequencies, as shown in Figure 10. Effectively, the broadband absorption is enhanced relative to a strictly non-graded material. These results are in agreement with the previous discussion of the Figure 9c,f.

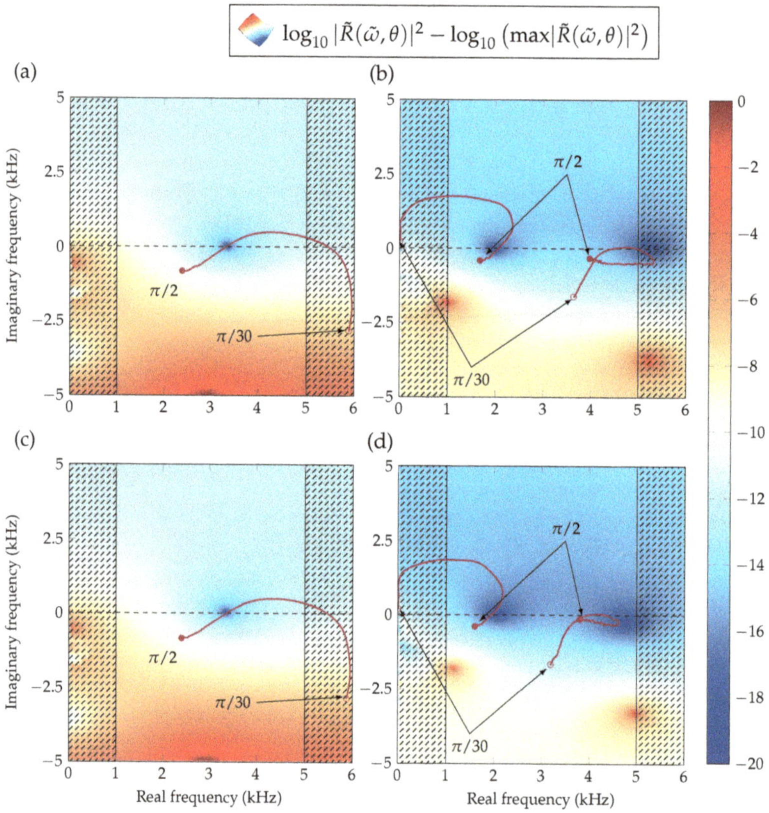

Figure 10. (Color online) Complex frequency plane representation of the reflection coefficient. The amplitude is normalized and reads $\log_{10}|\tilde{R}(\tilde{\omega},\theta)|^2 - \log_{10}\left(\max|\tilde{R}(\tilde{\omega},\theta)|^2\right)$ as a function of f_R and f_I. The maps are plotted at the incidence angle for which the first zero is on the real frequency axis in all optimization cases, $\bar{J}^{non-grad}$ (**a**), $\bar{J}^{grad}$ (**b**) and $J_{dif}^{non-grad}$ (**c**) , J_{dif}^{grad} (**d**). The paths of the complex zeros of reflection are displayed in solid purple, between $\theta_{min} = \pi/30$ (○) and $\theta_{max} = \pi/2$ (•).

6. Conclusions

In this work, an optimization technique is developed and applied to anisotropic and graded structures. Acoustic wave absorption can therefore be fine-tuned in porous materials, and designed for specific angular and frequency ranges. The proposed optimization routine relies on a database of 100 anisotropic porous unit cells, and provides the physical properties of the desired porous material presenting target absorption features. The effective properties, defined by the mass density tensor and bulk modulus, are graded, and thus macro-modulated along the depth of the porous material. Instead of running the minimization for all JCAL parameters of the equivalent fluid, the geometric parameters are preferred. The average and diffuse field absorption coefficients are considered, and lead to different geometric profiles for rigidly backed layers of finite thickness. While the optimization of the diffuse field absorption coefficient is weighted angularly, reducing the effects of the absorption at normal and grazing incidences, we have seen that using the average absorption can be useful to obtain larger values of absorption at the normal and grazing angles.

Sub-wavelength and broadband absorption objective functions are considered between 1 kHz and 5 kHz for a fixed thickness of 25 mm. From numerical results emerge the benefits of anisotropic and graded effective properties. They are visualized through the use of both the angular and the complex frequency maps, that reveal how the absorption coefficient is maximized. The resulting materials tend to approach omnidirectional absorption, by exploiting the database of anisotropic cells and graded properties. Moreover, as it has previously been found for normal incidence, the graded materials can strengthen the absorption properties by making use of additional resonances, produced by the graded properties of the system.

However, such a procedure is non-exhaustive, since a choice of unit cell and geometric parameters should first be made. It leads to optimized non-unique solutions for a given micro-structure.

Although the present numerical investigation gives promising results, the main hurdle resides in manufacturing such micro-structures. The manufacturing techniques can be difficult to apply for such designs, but the strength of the proposed method is its use on the database. Recent printing techniques can be put into use, but remain limited in terms of surface roughness and overall accuracy at the microscopic scale. In this way, the community is moving towards characterization and imaging techniques, so that surface roughness, and overall discrepancies of the manufactured medium can hence be accounted for, and be quantitatively estimated.

Author Contributions: Conceptualization, T.C., V.R.-G., J.-P.G., G.G., M.E., and J.R.; methodology, T.C., J.B., V.R.-G., J.-P.G., and G.G.; software, T.C., and J.B.; validation, T.C., J.-P.G., and J.B.; formal analysis, T.C., V.R.-G., J.-P.G., G.G., M.E., and J.R.; investigation, T.C., V.R.-G., J.-P.G., G.G., M.E., and J.R.; resources, T.C., J.B., V.R.-G., J.-P.G., and G.G.; data curation, T.C., J.B., V.R.-G., J.-P.G., and G.G.; writing—original draft preparation, T.C.; writing—review and editing, T.C., J.B., V.R.-G., J.-P.G., and G.G.; visualization, T.C.; supervision, V.R.-G., J.-P.G., G.G., M.E., and J.R.; project administration, G.G.; funding acquisition, G.G. All authors have read and agreed to the published version of the manuscript.

Funding: This research was funded by ANR Chaire industrielle MACIA (ANR-16-CHIN-0002).

Acknowledgments: The authors gratefully acknowledge the support from Dr. L. Schwan, for his impactful contribution regarding the numerical implementation and insight on physical interpretation.

Conflicts of Interest: The authors declare no conflict of interest.

Abbreviations

The following abbreviations are used in this manuscript:

TMM	Transfer matrix method
FEM	Finite elements method
PS	Peano series
WSTGF	Wave-splitting transfer Green functions
PCHIP	Piecewise cubic Hermite interpolating polynomial
JCAL	Johnson–Champoux–Allard–Lafarge

References

1. Zwikker, C.; Kosten, C.W. *Sound Absorbing Materials*; Elsevier: New York, NY, USA, 1949.
2. Attenborough, K. Acoustical characteristics of porous materials. *Phys. Rep.* **1982**, *82*, 179–227. [CrossRef]
3. Eijk, J.V.D.; Kosten, C.W.; Kok, W. Sound absorption by porous materials I. *Appl. Sci. Res. Sect. B* **1950**, *1*, 50–62. [CrossRef]
4. Paris, E. On the coefficient of sound-absorption measured by the reverberation method. *Lond. Edinb. Dublin Philos. Mag. J. Sci.* **1928**, *5*, 489–497. [CrossRef]
5. Ingard, K.U. *Noise Reduction Analysis*; Physics Series; Jones and Bartlett Publ: Sudbury, MA, USA, 2010.
6. Cox, T.J.; D'Antonio, P. *Acoustic Absorbers and Diffusers: Theory, Design And Application*, 3rd ed.; CRC Press, Taylor & Francis: Boca Raton, FL, USA, 2017.
7. Pierce, A.D. *Acoustics: An Introduction to Its Physical Principles and Applications*; Springer International Publishing: Cham, Switzerland, 2019. [CrossRef]
8. Allard, J.F. *Propagation of Sound in Porous Media: Modelling Sound Absorbing Materials*; Elsevier Applied Science: London, UK; New York, NY, USA, 1993.
9. Lafarge, D.; Lemarinier, P.; Allard, J.F.; Tarnow, V. Dynamic compressibility of air in porous structures at audible frequencies. *J. Acoust. Soc. Am.* **1997**, *102*, 1995–2006. [CrossRef]

10. Auriault, J.L.; Boutin, C.; Geindreau, C. *Homogenization of Coupled Phenomena in Heterogenous Media*; John Wiley & Sons: London, UK; Hoboken, NJ, USA, 2009.
11. Tarnow, V. Measured anisotropic air flow resistivity and sound attenuation of glass wool. *J. Acoust. Soc. Am.* **2002**, *111*, 2735–2739. [CrossRef]
12. Van der Kelen, C.; Göransson, P. Identification of the full anisotropic flow resistivity tensor for multiple glass wool and melamine foam samples. *J. Acoust. Soc. Am.* **2013**, *134*, 4659–4669. [CrossRef] [PubMed]
13. Terroir, A.; Schwan, L.; Cavalieri, T.; Romero-García, V.; Gabard, G.; Groby, J.P. General method to retrieve all effective acoustic properties of fully-anisotropic fluid materials in three dimensional space. *J. Appl. Phys.* **2019**, *125*, 025114. [CrossRef]
14. Boulvert, J.; Cavalieri, T.; Costa-Baptista, J.; Schwan, L.; Romero-García, V.; Gabard, G.; Fotsing, E.R.; Ross, A.; Mardjono, J.; Groby, J.P. Optimally graded porous material for broadband perfect absorption of sound. *J. Appl. Phys.* **2019**, *126*, 175101. [CrossRef]
15. Zieliński, T.G.; Venegas, R.; Perrot, C.; Červenka, M.; Chevillotte, F.; Attenborough, K. Benchmarks for microstructure-based modelling of sound absorbing rigid-frame porous media. *J. Sound Vib.* **2020**, *483*, 115441. [CrossRef]
16. Krueger, R.; Ochs, R. A green's function approach to the determination of internal fields. *Wave Motion* **1989**, *11*, 525–543. [CrossRef]
17. Borzdov, G.N. Frequency domain wave-splitting techniques for plane stratified bianisotropic media. *J. Math. Phys.* **1997**, *38*, 6328–6366. [CrossRef]
18. Shuvalov, A.; Poncelet, O.; Deschamps, M. General formalism for plane guided waves in transversely inhomogeneous anisotropic plates. *Wave Motion* **2004**, *40*, 413–426. [CrossRef]
19. Baron, C. Matricant Peano Series Development to Study Elastic Waves Propagation in Continuously Varying Properties Materials. Ph.D. Thesis, Université de Bordeaux, Bordeaux, France, 2005.
20. Gautier, G.; Kelders, L.; Groby, J.P.; Dazel, O.; De Ryck, L.; Leclaire, P. Propagation of acoustic waves in a one-dimensional macroscopically inhomogeneous poroelastic material. *J. Acoust. Soc. Am.* **2011**, *130*, 1390–1398. [CrossRef]
21. Geslain, A.; Groby, J.P.; Dazel, O.; Mahasaranon, S.; Horoshenkov, K.V.; Khan, A. An application of the Peano series expansion to predict sound propagation in materials with continuous pore stratification. *J. Acoust. Soc. Am.* **2012**, *132*, 208–215. [CrossRef]
22. Cavalieri, T.; Boulvert, J.; Schwan, L.; Gabard, G.; Romero-Garcìa, V.; Groby, J.P.; Escouflaire, M.; Mardjono, J. Acoustic wave propagation in effective graded fully anisotropic fluid layers. *J. Acoust. Soc. Am.* **2019**, *146*, 3400–3408. [CrossRef]
23. Pease, M.C. *Methods of Matrix Algebra*; Academic Press: New York , NY, USA, 1965.
24. Brouard, B.; Lafarge, D.; Allard, J.F. A general method of modelling sound propagation in layered media. *J. Sound Vib.* **1995**, *183*, 129–142. [CrossRef]
25. Teperik, T.V.; García de Abajo, F.J.; Borisov, A.G.; Abdelsalam, M.; Bartlett, P.N.; Sugawara, Y.; Baumberg, J.J. Omnidirectional absorption in nanostructured metal surfaces. *Nat. Photonics* **2008**, *2*, 299–301. [CrossRef]
26. Fritsch, F.N.; Carlson, R.E. Monotone Piecewise Cubic Interpolation. *SIAM J. Numer. Anal.* **1980**, *17*, 238–246. [CrossRef]
27. Darcy, H. *Les Fontaines Publiques de la ville de Dijon: Exposition et application des principes à suivre et des formules à employer dans les questions de distribution d'eau*; Hachette Livre BNF: Paris, France, 1856.
28. Nelder, J.A.; Mead, R. A Simplex Method for Function Minimization. *Comput. J.* **1965**, *7*, 308–313. [CrossRef]
29. Press, W.H. (Ed.) *Numerical Recipes: The Art of Scientific Computing*, 3rd ed.; Cambridge University Press: Cambridge, UK; New York, NY, USA, 2007.
30. Romero-García, V.; Theocharis, G.; Richoux, O.; Pagneux, V. Use of complex frequency plane to design broadband and sub-wavelength absorbers. *J. Acoust. Soc. Am.* **2016**, *139*, 3395–3403. [CrossRef]
31. Groby, J.P.; Pommier, R.; Aurégan, Y. Use of slow sound to design perfect and broadband passive sound absorbing materials. *J. Acoust. Soc. Am.* **2016**, *139*, 1660–1671. [CrossRef] [PubMed]

32. Luk, T.S.; Campione, S.; Kim, I.; Feng, S.; Jun, Y.C.; Liu, S.; Wright, J.B.; Brener, I.; Catrysse, P.B.; Fan, S.; et al. Directional perfect absorption using deep subwavelength low-permittivity films. *Phys. Rev. B* **2014**, *90*, 085411. [CrossRef]
33. Romero-García, V.; Theocharis, G.; Richoux, O.; Merkel, A.; Tournat, V.; Pagneux, V. Perfect and broadband acoustic absorption by critically coupled sub-wavelength resonators. *Sci. Rep.* **2016**, *6*, 19519. [CrossRef] [PubMed]

materials

MDPI

Article

An Investigation on the Sound Absorption Performance of Granular Molecular Sieves under Room Temperature and Pressure

Bing Zhou [1], Jiangong Zhang [1,*], Xin Li [2] and Bilong Liu [2,*]

[1] State Key Laboratory of Power Grid Environmental Protection, China Electric Power Research Institute, Wuhan 430074, China; zhoubing2@epri.sgcc.com.cn

[2] School of Mechanical and Automobile Engineering, Qingdao University of Technology, Qingdao 266520, China; jz03-4lx@163.com

* Correspondence: zhangjiangong@epri.sgcc.com.cn (J.Z.); liubilong@qut.edu.cn (B.L); Tel.: +86-185-6395-0863 (J.Z.); +86-136-8337-1062 (B.L)

Received: 29 February 2020; Accepted: 16 April 2020; Published: 20 April 2020

check for updates

Abstract: The sound absorption of granular silica-aluminate molecular sieve pellets was investigated in this paper. The absorption coefficients of molecular sieve pellets with different pore sizes, pellet sizes, and layer thicknesses were measured through impedance tubes under room temperature and pressure conditions. The effects of pore size, pellet size, layer thickness were compared and explained. The comparisons show that at room temperature and pressure, the sound absorption of molecular sieve pellets is not a result of the crystalline structure, but rather it mainly changes with the pellet size and layer thickness. In addition, the five non-acoustical parameters of molecular sieve pellets were obtained by an inverse characterization method based on impedance tube measurements. The measurement by impedance tubes is in good agreement with the calculation of Johnson-Champoux-Allard (JCA) model, proving that the JCA model can be effectively used to predict the sound absorption of molecular sieve pellets.

Keywords: molecular sieve pellets; sound absorption; impedance tube

---

1. Introduction

Porous materials are the most widely used sound absorbing materials in engineering. The obvious sound absorption of porous materials lies in the fact that their internal structure has numerous tiny pores that are interlinked with each other and connected to the outside on their surfaces. During the process of sound wave propagation, friction occurs between the moving air in the pores and the solid skeleton of the pore wall, and viscous loss is caused by the viscosity of airflow within the materials. Due to the viscosity and heat conduction effect of the air, sound energy is converted into internal energy and then dissipated, thus achieving sound absorption [1].

Porous sound absorbing materials may be classified into fibrous, cellular and granular [2,3]. Granular materials are considered as rigid porous materials due to their fixed framework, and their sound energy dissipation mainly depends on the air viscosity in the gaps of the granular materials and the heat transfer between the air and the granular material surface. Meanwhile, the scattering of granules will also affect the absorption of sound energy in the material. References [4–10] studied the sound absorption of some granular porous materials such as rubber chips, gravel, perlite, aerogel and concrete. The results show that these granular porous materials play a role in sound absorption and noise reduction of road noise barriers and buildings. Molecular sieves are a kind of granular porous material that have been widely used as air adsorbents and desiccants in many applications [11,12] Quite recently, granular molecular sieves have been used as sound absorption materials inside the back cavities of miniature loudspeakers for cellphones [13,14]. It is considered that the air absorption mechanism increases the acoustic compliance of a volume of molecular sieves, which significantly reduces the first resonance frequency of the coupled membrane-cavity system and thus improves the sound performance of loudspeakers. For underwater acoustics, a damping material consisting of Rho c-carbon fiber matrix and molecular sieve pellets was reported, and experimental results showed that the addition of molecular sieve pellets improved the sound insulation [15]. On the topic of the sound absorption characteristics of molecular sieves as porous materials, however, there are few reports making this worth investigating in some detail.

Molecular sieves, also known as zeolites, belong to the aluminosilicate compounds, a sort of crystalline porous material. The corresponding crystallographic structure is formed by tetrahedrons of AlO_4 and SiO_4, which are the basic components of different zeolites. The crystalline structure of molecular sieves is a three-dimensional porous system that precisely adjusts the pore size to allow molecules smaller than this pore size to be adsorbed and to exclude larger molecules [11,12], hence the name "molecular sieve". Molecular sieves are usually divided into A, X, Y, M and ZSM types according to their ratio of aluminum to silicon. Generally speaking, the higher the Al-Si ratio of a zeolite is, the better its stability is. For example, the initial damage temperature of the structure of NaA ($SiO_2/Al_2O_3 = 2$) and NaX ($SiO_2/Al_2O_3 = 2.5$) is approximately 660 °C. Molecular sieves with different crystal structures are usually classified by the pore diameter, such as 3Å, 4Å and 5Å, where 1Å = 0.1 nm. 3Å molecular sieves have effective adsorption selectivity and are suitable for the insulating glass desiccant and refrigerant desiccant applications; 5Å molecular sieves have high adsorption and fast adsorption speed, and are suitable for various gas pressure swing adsorption devices for oxygen, hydrogen and carbon dioxide production. The appearance of molecular sieves is usually as a powder, strips or granular. The size range of molecular sieve pellets is generally 0.4–0.8, 1.6–2.5 or 3–5 mm. For convenience, we will uniformly refer to them as micro, small and large pellets in this paper.

The acoustic properties of porous materials are mainly focused on the propagation of sound waves through their interior. Zwikker and Kosten [16] pointed out that for isotropic homogeneous media, the propagation of sound waves can be described by the characteristic impedance and propagation constant of the materials. Due to the complexity of the factors affecting the sound absorption of porous materials, empirical models are generally used to estimate the sound absorption performance. Delany-Bazley [17] obtained an empirical formula to describe the characteristic impedance and wave number of porous materials through a large number of measurements on fibrous materials. This empirical model is very successful since it requires only the static flow resistivity of materials. The model, however, is not applicable to the prediction of sound absorption in wide frequency band. Miki [18] further modified the above formula to make it applicable to a wider frequency range. In addition to the empirical formula above, many scholars proposed a semi-phenomenological model to describe the equivalent dynamic density and equivalent volume modulus of fluids in rigid skeletal porous materials [19–25]. In 1987, Johnson et al. [19] proposed a semi-phenomenological model to describe the complex density of porous materials with arbitrary pore shapes and static skeletons. Taking into account the viscosity-inertia effect, four parameters that are static flow resistivity, porosity, tortuosity and viscosity characteristic length, were used to calculate dynamic density. In 1991, Champoux and Allard [20] further considered the heat conduction effect and gave the expression of dynamic volume modulus of porous materials with the same property. Five parameters, which are static flow resistivity, porosity, tortuosity, viscous characteristic length and thermal characteristic length, were used to calculate the equivalent fluid dynamic density and equivalent fluid dynamic volume modulus of the porous material model. Johnson-Champoux-Allard (JCA) model is a generalized model suitable for the wide-band propagation of sound waves in porous materials [21–23] Non-acoustic parameters of porous materials can be obtained through direct measurement or inverse acoustical characterization [26,27]. Direct measurement requires special laboratory equipment. Static flow resistance and porosity can be directly measured by standard technology, while tortuosity and viscous characteristic length, thermal characteristic length can be measured by ultrasonic technology, which is difficult to implement under conventional laboratory conditions, and the measurement accuracy cannot easily be guaranteed. Inverse acoustic characterization is an alternative to direct measurements. Once the acoustic impedance or absorption coefficient of the material is known, the relevant parameters of the material can be calculated using the inverse strategy. In [7,26,28,29] the non-acoustic parameters of porous materials are evaluated using inverse acoustic characterization, and the results show that this method is robust and reliable. Here, taking the minimum difference between the sound absorption coefficient measured by the impedance tube and calculated by the JCA model as the objective function, five non-acoustic parameters of molecular sieve are obtained by using simulated annealing method.

The paper is organized as follows: In Section 2, sound absorption measurements of different molecular sieve pellets samples with different thickness and different pellet size are reported and analyzed; and in Section 3, JCA model is adopted to describe the acoustical properties of molecular sieve pellets. Five non-acoustic parameters of the molecular sieve pellets are calculated by simulated annealing. Moreover, the reliability of JCA model for predicting the sound absorption of molecular sieve pellets is verified. Finally, the conclusions are summarized in Section 4.

2. Acoustic Measurement of Molecular Sieve Pellets

2.1. Experimental Equipment and Samples

The purpose of this paper is to investigate the sound absorption characteristics of molecular sieves at room temperature and pressure. 3Å and 5Å molecular sieve pellets were chosen for the measurements carried out in SW impedance tubes according to ISO 10534-2 [30]. Based on the transfer function method, the normal incident sound absorption coefficient and impedance in the 63–6300 Hz range are measured by SW422 and SW477 impedance tubes, of which the SW422 has an inner diameter of 100 mm and is used for 63–1600 Hz and the SW477 has 30 mm and is used for 1000–6300 Hz measurements. As for the impedance tube test system in Figure 1, a BSWA PA50 power amplifier drives the speaker to amplify the sound, the microphones located at positions 1 and 2 collect the sound pressure of the incident and reflected waves, and a MC3242 data acquisition system analyses the sound pressure signals. The reflection factor of the sound waves is calculated by the transfer function of incident and reflected waves in sound field. Then the normal incident coefficient and surface acoustic impedance are calculated. The surface acoustic impedance and impedance are the real and imaginary parts of the surface acoustic impedance, respectively.

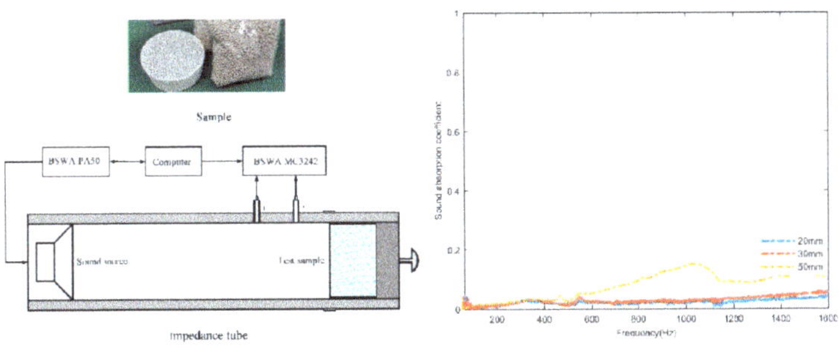

Figure 1. Experiment setup. On the left are the sample and the impedance tube, and on the right is the absorption coefficient of the cylindrical box.

A perforated cylindrical box is designed to fill with loose molecular sieve pellets. The sound absorption of the perforated cylinder is very low and can be ignored here. For example, a perforated cylinder with an aperture of 1.5 mm and a perforation ratio of 39.6% is used, and its measurement of sound absorption coefficient with cylinder depths of 20, 30 and 50mm is shown on the right of Figure 1. It should be noted that, in the stacking state, the void ratio of bulk granular materials is described by the packing density and apparent density and is related to the sound absorption of granular materials. The void ratio is inversely proportional to the packing density of materials. Molecular sieves pellets described here are randomly closely packed in the specimen cylinder in a natural state without additional force or adhesive compaction. In the following sections, the measured normal sound absorption coefficient and surface acoustic impedance of molecular sieve pellets of different pore size and different pellet sizes, and different thickness are given and compared.

2.2. Measured Sound Absorption of Molecular Sieve Pellets with Different Pore Size

In order to find out the effect of pore size on the sound absorption of molecular sieve pellets under room temperature and pressure, two types of samples with pore sizes of 3Å and 5Å are selected for sound absorption comparison. The characteristics of the molecular sieve pellet samples used are listed in Table 1. Figure 2a–d show the sound absorption curves of molecular sieve pellets of the same layer thick and pellet size.

Table 1. Characteristics of the molecular sieve pellet samples with different pore sizes.

3Å			5Å		
Number	**Pellet Size (mm)**	**Thickness (mm)**	**Number**	**Pellet Size (mm)**	**Thickness (mm)**
#1	3–5	30	#1	3–5	30
#2	3–5	50	#2	3–5	50
#3	1.6–2.5	30	#3	1.6–2.5	30
#4	1.6–2.5	50	#4	1.6–2.5	50

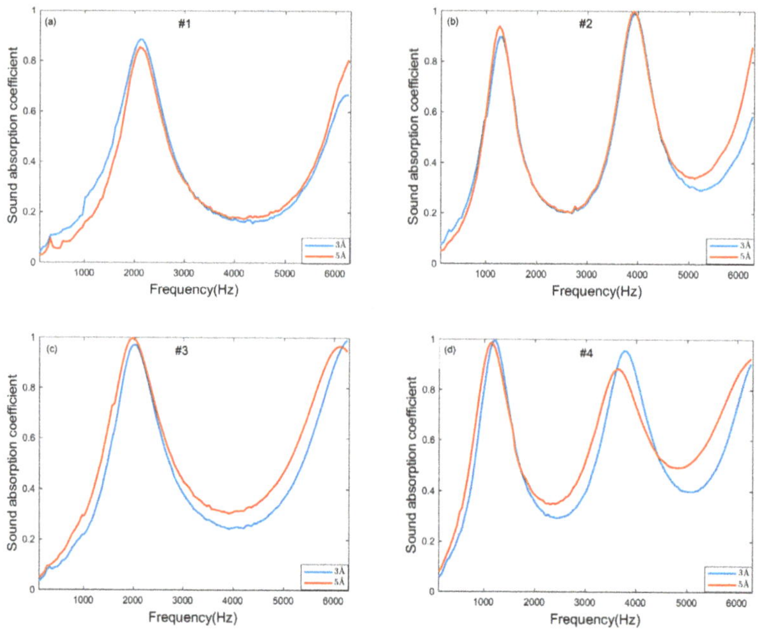

Figure 2. Sound absorption measurement of molecular sieve pellets of different pore size. Sound absorption coefficient of 3Å and 5Å with (**a**) thickness of 30 mm and pellet size of 3–5 mm; (**b**) thickness of 50 mm and pellet size of 3–5 mm; (**c**) thickness of 30 mm and pellet size of 1.6–2.5 mm; (**d**) thickness of 50 mm and pellet size of 1.6–2.5 mm.

The sound absorption curves of 3Å and 5Å are similar in shape and almost identical. The slight differences in some frequency bands may be caused by installation or measurement errors. Under room temperature and pressure, "sieve molecules" does not appear due to its own strengthening stability. That is to say, the pore size has no obvious effect on the sound absorption of molecular sieve pellets. The sound absorption of molecular sieve pellets is mainly related to the macroscopic size of pellets and is less related to the pore size. Therefore, molecular sieve pellets are regarded as rigid porous materials, and sound energy is mainly consumed by the air viscosity between pellets and heat transfer between air and pellets surface. In addition, a low sound absorption coefficient valley appears on each sound absorption curve, which obviously affects the wide-band sound absorption of molecular sieve pellets. This is because the sound absorption of molecular sieve pellets is related to flow resistance, porosity and structure factor, and finally attributed to the pellets gap. When the gap between pellets is too large or too small, the air between the pellets cannot be sufficiently frictional and heat transfer during the propagation of sound waves.

2.3. Measured Sound Absorption of Molecular Sieve Pellets with Different Thickness

To illustrate the effect of layer thickness on the sound absorption of molecular sieve pellets, the measured sound absorption coefficient for the eight samples given in Table 2 are plotted in Figure 3. As shown in Figure 3a,b, as the thickness of the molecular sieve pellets increases, the absorption peak tends to the low frequency and as well the effective absorption bandwidth becomes narrower. As the thickness of the layer increases, the gap distribution becomes more complex and the gap channel becomes longer. In this way, during the acoustic wave propagation process, the number of collisions between pellets increases and more energy is consumed, and the flow volume in the gap increases, so the natural frequency of molecular sieve pellets decreases and the sound absorption moves to a low frequency.

Table 2. Samples of molecular sieve pellets with different layer thickness.

3Å			5Å		
Number	**Pellet Size (mm)**	**Thickness (mm)**	**Number**	**Pellet Size (mm)**	**Thickness (mm)**
#1	3–5	20	#1	1.6–2.5	20
#2	3–5	30	#2	1.6–2.5	30
#3	3–5	40	#3	1.6–2.5	40

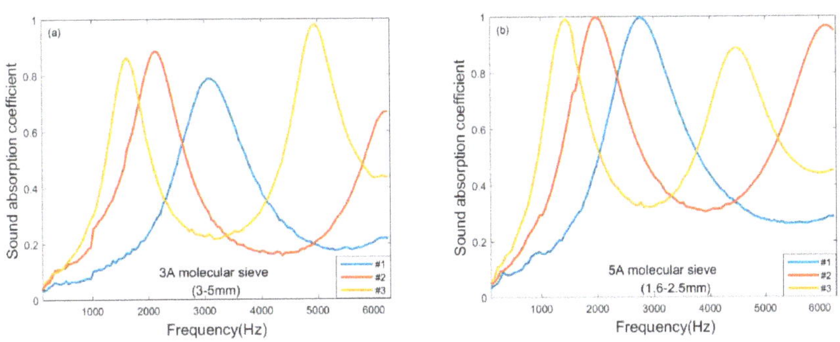

Figure 3. Sound absorption measurement of molecular sieve pellets with different thickness: (**a**) 3Å with pellet size of 3–5 mm; (**b**) 5Å with pellet size of 1.6–2.5 mm.

2.4. Measured Sound Absorption of Molecular Sieve Pellets with Different Pellet Sizes

Three different pellet size molecular sieve pellets selected for sound absorption comparison are illustrated in Figure 4, where the (a1) and (b1) are the sound absorption coefficient of the samples, (a2) and (b2) are the specific surface acoustic resistance of the samples, and (a3) and (b3) are the specific surface acoustic resistance of the samples. The pellet sizes are micro (0.4–0.8 mm), small (1.6–2.5 mm), and large (3–5 mm), as shown in Table 3. It is evident that as the pellet size decreases, the sound absorption peaks of the molecular sieve pellets move to a low frequency, and the effective sound absorption bandwidth increases significantly. Compared with the large and small samples, the sound absorption coefficient of the micro sample is smaller and flat, as shown in (a1) and (b1). The sound absorption coefficient of the micro sample is greater than 0.6 in the bandwidth of 500–6300 Hz. When the molecular sieve pellets are smaller, the pellet gap is smaller and denser, and the fluid resistance in the gap increases, thus more sound energy is lost. This characteristic is clearly reflected from the measured specific surface acoustic resistances in (a2) and (b2), where the micro sample has the largest acoustic resistance excluding the frequency at resonance. Also the measured specific surface acoustic reactance of the micro sample are the nearest zero in the whole frequency range, therefore the smallest samples have the benefit for broadband sound absorption. These results imply that appropriate pellet size should be taken into account carefully to have a satisfactory sound absorption.

Table 3. Samples of molecular sieve pellets with different pellet sizes.

5Å					
Number	**Pellet Size (mm)**	**Thickness (mm)**	**Number**	**Pellet Size (mm)**	**Thickness (mm)**
#11	0.4–0.8	30	#21	0.4–0.8	50
#12	1.6–2.5	30	#22	1.6–2.5	50
#13	3–5	30	#23	3–5	50

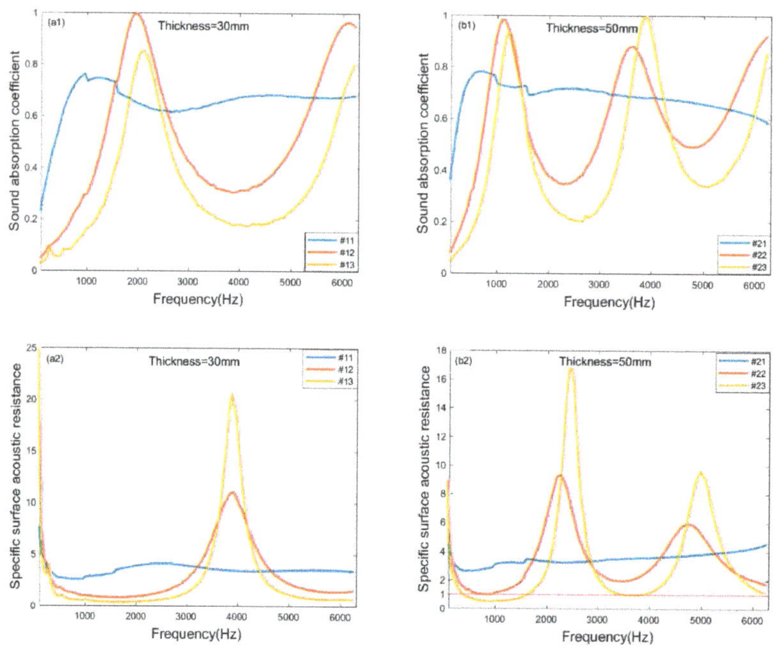

Figure 4. *Cont.*

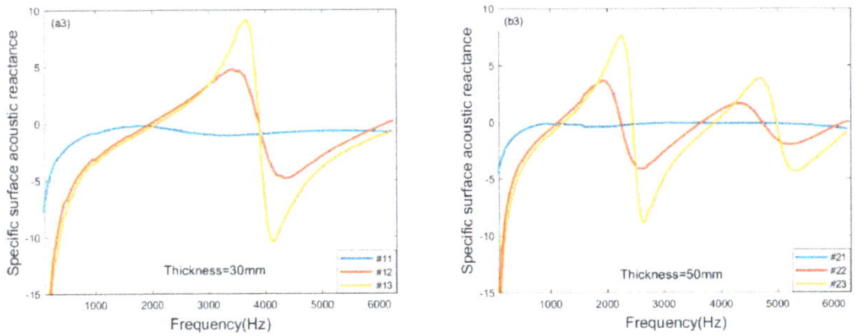

Figure 4. Sound absorption measurement of molecular sieve pellets with different pellet sizes. Sound absorption coefficients of 5Å with thickness of (**a1**) 30 mm; (**b1**) 50 mm; Specific surface acoustic resistance of 5Å with thickness of (**a2**) 30 mm; (**b2**) 50 mm; Specific surface acoustic reactance of 5Å with thickness of (**a3**) 30 mm; (**b3**) 50 mm.

3. JCA model and Inverse Parameters Estimation

3.1. Theoretical Model

JCA model is a generalized model for sound propagation in porous materials over a wide range of frequencies, which includes five non-acoustic parameters of porous materials: the static flow resistance σ, porosity ϕ, the tortuosity α_∞, the viscous characteristic length Λ, and the thermal characteristic length $\Lambda\prime$. In JCA model, the equivalent dynamic density $\rho_e(\omega)$ and the equivalent dynamic volume modulus $K_e(\omega)$ are expressed as [20]:

$$\rho_e(\omega) = \rho_0\alpha_\infty\left(1 + \frac{\phi\sigma}{j\omega\rho_0\alpha_\infty}\left(1 + j\frac{4\omega\rho_0\eta{\alpha_\infty}^2}{(\sigma\phi\Lambda)^2}\right)^{1/2}\right) \tag{1}$$

$$K_e(\omega) = \frac{\gamma P_0}{\gamma - (\gamma-1)\left(1 + \frac{8\eta}{j\omega B^2\Lambda\prime^2\rho_0}\left(1 + j\frac{\omega B^2\rho_0\Lambda\prime^2}{16\eta}\right)^{1/2}\right)^{-1}} \tag{2}$$

$$\Lambda = \frac{1}{c}\left[\frac{8\alpha_\infty\eta}{\sigma\phi}\right]^{1/2} \tag{3}$$

$$\Lambda' = \frac{1}{c'}\left[\frac{8\alpha_\infty\eta}{\sigma\phi}\right]^{1/2}, \tag{4}$$

where P_0 is the ambient atmospheric pressure, ω is the angular frequency, ρ_0, η, γ and B^2 are the density, dynamic viscosity, specific heat ratio and Prandtl number of the saturating air, respectively. In Equations (3) and (4), c and $c\prime$ are pore shape parameters related to the viscous and thermal dissipation, respectively.

Based on the equivalent fluid theory, the characteristic acoustic impedance Z_c and complex wave number k of rigid porous materials are expressed as:

$$Z_c = \frac{1}{\phi}\sqrt{\rho_e(\omega)gK_e(\omega)} \tag{5}$$

$$k = \omega\sqrt{\frac{\rho_e}{K_e}}. \tag{6}$$

Therefore, according to Zwicker and Kosten theory, the surface acoustic impedance Z_s of porous materials backed by rigid walls is expressed as:

$$Z_s = Z_c \coth(kL). \tag{7}$$

Then, the normal incidence sound absorption coefficient can be calculated as:

$$\alpha = 1 - \left| \frac{Z_s - \rho_0 c_0}{Z_s + \rho_0 c_0} \right|^2, \tag{8}$$

where, L is the thickness of the porous material layer, $\rho_0 c_0$ is the characteristic impedance of the air.

3.2. Inversion of Non-Acoustic Parameters

Simulated annealing is an optimization algorithm based on Monte Carlo iteration. Because of its global optimization and strong robustness, it is widely used in the optimization of multi-objective parameters in acoustic materials and structures [26,31–34]. In this paper, the inverse calculation of five non-acoustic parameters is transformed into the solution of a set of optimal values. The process is as follows: firstly, the difference between the sound absorption coefficient measured in the impedance tube and calculated by JCA model in a certain frequency band is taken as the optimization objective function, and then a set of optimization values are obtained by global optimization calculation using simulated annealing method. The objective function is as follows:

$$F(\alpha) = \sum_i \left| \alpha_{Measured}(f_i) - \alpha_{JCA}(f_i) \right|, \tag{9}$$

where f_i is the frequency point within the specified frequency range, $\alpha_{Measured}(f_i)$ is the sound absorption coefficient at the frequency point f_i measured by the impedance tube, $\alpha_{JCA}(f_i)$ is the sound absorption coefficient calculated by JCA model.

Meanwhile, according to the analysis of five non-acoustic parameters in literature [26], the optimization interval of five parameters in this paper is set as:

$$\begin{cases} \sigma \in \left[\ 1000 \quad 10,000 \ \right] \text{Pa·s·m}^{-2} \\ \phi \in \left[\ 0.1 \quad 0.9 \ \right] \\ \alpha_\infty \in \left[\ 1 \quad 4 \ \right] \\ c \in \left[\ 0.3 \quad 3.3 \ \right] \\ c' \in \left[\ 0.3 \quad 3 \ \right] \end{cases}. \tag{10}$$

Using the above inversion strategy, the calculated non-acoustic parameters of samples are listed in Table 4. The sound absorption coefficient measured using impedance tubes and estimated by substituting the inversion values into the JCA model are respectively drawn in Figure 5. It can be found that in the frequency range of 63-6300Hz, the measured curve is almost identical to the estimated curve.

Table 4. Inverse non-acoustic parameters estimated by simulated annealing.

Sample	Pellet Size (mm)	Thickness (mm)	σ (Pa·s·m^{-2})	ϕ (%)	α_∞	Λ (μm)	Λ' (μm)	c	c'
3Å	3–5	40	4535	39.39	1.62	360.32	540.77	1	0.66
5Å	1.6–2.5	50	4503	44.62	1.68	216.20	345.93	1.60	1

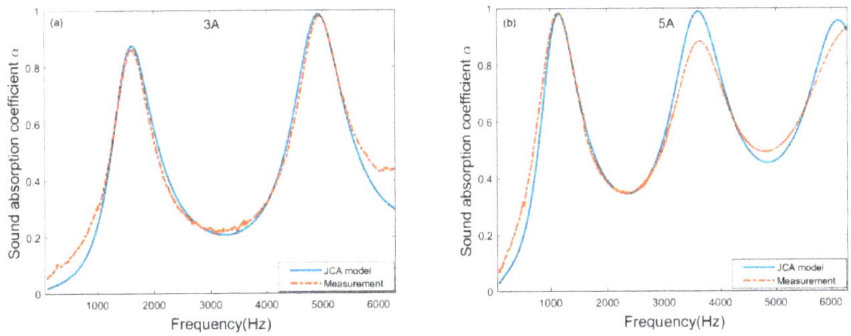

Figure 5. Comparison between measurement by impedance tube and estimation through inversion of JCA model. (**a**) 3Å molecular sieve pellets with thickness of 40 mm; (**b**) 5Å molecular sieve pellets with thickness of 50 mm.

3.3. Validation

New samples with different thickness listed in Table 5 are chosen to verify that the inversion parameters estimated in Table 4. The sound absorption coefficients of these new samples measured and predicted using the inversion parameters are shown in Figure 6. The prediction of sample #1 and #2 is based on the inversion parameters of sample 3Å in Table 4 and the sample #3 and #4 is according to sample 5Å in Table 4. It is evident that there is good agreement between the predicted and measured sound absorption coefficients for these samples. Therefore, the inversion parameters appear to be independent of the thickness of molecular sieve pellets. This implies that the JCA model can be used as a reliable model to predict the sound absorption performance of molecular sieve pellets, and the five non-acoustic parameters obtained from one sample by inverse method based on the JCA model and simulated annealing are reliable to be used for the prediction of the sound absorption coefficient of molecular sieve pellets with different thickness.

Table 5. Samples of molecular sieve pellets for prediction.

3Å			5Å		
Number	**Pellet Size (mm)**	**Thickness (mm)**	**Number**	**Pellet Size (mm)**	**Thickness (mm)**
#1	3–5	20	#3	1.6–2.5	30
#2	3–5	50	#4	1.6–2.5	40

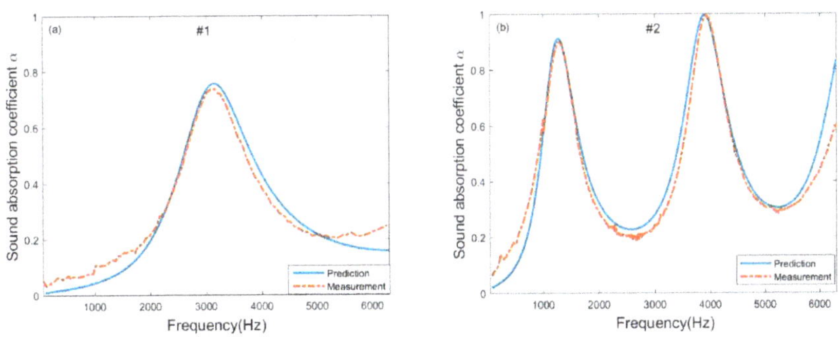

Figure 6. *Cont.*

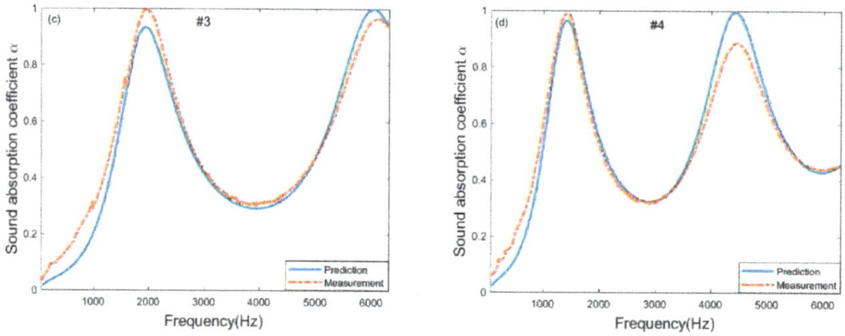

Figure 6. Comparison between measurement and prediction for the samples listed in using the inversion parameters estimated in **(a)** new sample #1; **(b)** new sample #2; **(c)** new sample #3; **(d)** new sample #4.

4. Conclusions

Under the conditions of room temperature and pressure, the sound absorption performance of molecular sieve pellets was measured for comparisons. It is found that the sound absorption coefficients are very sensitive to the pellet size and the thickness of the molecular sieve pellets, but not sensitive to the pore size of the molecular sieve pellets, implying that the sound absorption is not caused by the crystalline structure. Molecular sieve pellets with optimized pellet size may result in satisfactory absorption in the broad band frequency range. Increasing the thickness of the molecular sieve pellets moves the sound absorption peaks to a lower frequency range. Moreover, it has been demonstrated that the JCA model is appropriate to describe the sound absorption performance of molecular sieve pellets. The five non-acoustic parameters obtained with one sample by the inverse method are reliable and can be used for the prediction of the sound absorption coefficient of molecular sieve pellets with different thicknesses. When the temperature and pressure are higher than those of room conditions, or under the condition of high sound intensity, the effect of the crystalline structure on the sound absorption has not been verified and is worth being investigated in the future.

Author Contributions: Conceptualization, B.L., B.Z. and J.Z.; methodology X.L.; Measurements, X.L., B.Z. and J.Z.; writing—original draft preparation X.L., B.Z. and B.L.; All authors have read and agreed to the published version of the manuscript.

Funding: Project supported by State Grid Corporation Science and Technology Project (GYW17201700090).

Acknowledgments: The authors gratefully acknowledge the support provided by the Taishan Scholar Program of Shandong.

Conflicts of Interest: The authors declare no conflict of interest.

References

1. Cao, L.; Qiuxia, F.; Yang, S.; Bin, D.; Jianyong, Y. Porous materials for sound absorption. *Compos. Commun.* **2018**, *10*, 25–35. [CrossRef]
2. Kidner, M.R.; Hansen, C.H. A comparison and review of theories of the acoustics of porous materials. *Int. J. Acoust. Vib.* **2008**, *13*, 112–119.
3. Arenas, J.P.; Crocker, M.J. Recent trends in porous sound-absorbing materials. *Sound Vibr.* **2010**, *44*, 12–18.
4. Pfretzschner, J.; Rodriguez, R.M. Acoustic properties of rubber crumbs. *Polym. Test.* **1999**, *18*, 81–92. [CrossRef]
5. Voronina, N.; Horoshenkov, K. A new empirical model for the acoustic properties of loose granular media. *Appl. Acoust.* **2003**, *64*, 415–432. [CrossRef]

6. Mohammed, G.; Benayad, C. Sound Absorption Measurements and Numerical Estimation of Non Acoustical Parameters of some Moroccan Raw Green Materials. In *Advanced Materials Research*; Trans Tech Publications: Bach, Switzerland, 2014; pp. 95–101.
7. Cobo, P.; Moraes, E.; Simón, F. Inverse estimation of the non-acoustical parameters of loose granular absorbers by Simulated Annealing. *Build. Environ.* **2015**, *94*, 859–866. [CrossRef]
8. Buratti, C.; Merli, F.; Moretti, E. Aerogel-based materials for building applications: Influence of granule size on thermal and acoustic performance. *Energy Build.* **2017**, *152*, 472–482. [CrossRef]
9. Wang, D.; Schacht, A.; Leng, Z.; Leng, C.; Kollmann, J.; Oeser, M. Effects of material composition on mechanical and acoustic performance of poroelastic road surface (PERS). *Constr. Build. Mater.* **2017**, *135*, 352–360. [CrossRef]
10. Oancea, I.; Bujoreanu, C.; Budescu, M.; Benchea, M.; Grădinaru, C.M. Considerations on sound absorption coefficient of sustainable concrete with different waste replacements. *J. Clean. Prod.* **2018**, *203*, 301–312. [CrossRef]
11. Davis, M.E.; Lobo, R.F. Zeolite and molecular sieve synthesis. *Chem. Mater.* **1992**, *4*, 756–768. [CrossRef]
12. Milton, R.M. Molecular Sieve Adsorbents. U.S. Patent 2882243A, 14 April 1959.
13. Bose, A.G. Loudspeaker System. U.S. Patent 3582553A, 1 June 1971.
14. Gao, Y.; Kang, X.; Li, J.; Zhang, F.; Kang, N.; Liu, H. Molecular Sieve, Sound Absorbing Material Using the Same, and Speaker. U.S. Patent 16/236336, 7 April 2019.
15. Philip, B.; Abraham, J.K.; Varadan, V.K.; Natarajan, V.; Jayakumari, V.G. Passive underwater acoustic damping materials with Rho-C rubber–carbon fiber and molecular sieves. *Smart Mater. Struct.* **2004**, *13*, N99–N104. [CrossRef]
16. Zwikker, C.; Kosten, C.W. Sound Absorbing Materials (F. Trendelenburg). *Zeitschrift Naturforschung Teil A* **1950**, *5*, 347.
17. Delany, M.E.; Bazley, E.N. Acoustical properties of fibrous absorbent materials. *Appl. Acoust.* **1970**, *3*, 105–116. [CrossRef]
18. Miki, Y. Acoustical properties of porous materials-Modifications of Delany-Bazley models. *J. Acoust. Soc. Jpn.* **1990**, *11*, 19–24. [CrossRef]
19. Johnson, D.L.; Koplik, J.; Dashen, R. Theory of dynamic permeability and tortuosity in fluid-saturated porous media. *J. Fluid Mech.* **1987**, *176*, 379–402. [CrossRef]
20. Champoux, Y.; Allard, J.F. Dynamic tortuosity and bulk modulus in air-saturated porous media. *J. Appl. Phys.* **1991**, *70*, 1975–1979. [CrossRef]
21. Champoux, Y.; Stinson, M.R.; Daigle, G.A. Air-based system for the measurement of porosity. *J. Acoust. Soc. Am.* **1991**, *89*, 910–916. [CrossRef]
22. Allard, J.F.; Champoux, Y. New empirical equations for sound propagation in rigid frame fibrous materials. *J. Acoust. Soc. Am.* **1992**, *91*, 3346–3353. [CrossRef]
23. Champoux, Y.; Stinson, M.R. On acoustical models for sound propagation in rigid frame porous materials and the influence of shape factors. *J. Acoust. Soc. Am.* **1992**, *92*, 1120–1131. [CrossRef]
24. Lafarge, D.; Allard, J.F.; Brouard, B.; Verhaegen, C.; Lauriks, W. Characteristic dimensions and prediction at high frequencies of the surface impedance of porous layers. *J. Acoust. Soc. Am.* **1993**, *93*, 2474–2478. [CrossRef]
25. Lafarge, D.; Lemarinier, P.; Allard, J.F.; Tarnow, V. Dynamic compressibility of air in porous structures at audible frequencies. *J. Acoust. Soc. Am.* **1997**, *102*, 1995–2006. [CrossRef]
26. Atalla, Y.; Panneton, R. Inverse acoustical characterization of open cell porous media using impedance tube measurements. *Can. Acoust.* **2005**, *33*, 11–24.
27. Bonfiglio, P.; Pompoli, F. Inversion problems for determining physical parameters of porous materials: Overview and comparison between different methods. *Acta Acust. United Acust.* **2013**, *99*, 341–351. [CrossRef]
28. Zhang, B.; Zhu, J. Inverse methods of determining the acoustical parameters of porous sound absorbing metallic materials. Proceedings of 22nd International Congress on Acoustics (ICA2016), Buenos Aires, Argentina, 5–9 September 2016.
29. Berardi, U.; Iannace, G. Predicting the sound absorption of natural materials: Best-fit inverse laws for the acoustic impedance and the propagation constant. *Appl. Acoust.* **2017**, *115*, 131–138. [CrossRef]
30. Chung, J.Y.; Blaser, D.A. Transfer function method of measuring in-duct acoustic properties. II. Experiment. *J. Acoust. Soc. Am.* **1998**, *68*, 914–921. [CrossRef]

31. Chang, Y.C.; Yeh, L.J.; Chiu, M.C. Optimization of constrained composite absorbers using simulated annealing. *Appl. Acoust.* **2005**, *66*, 341–352. [CrossRef]
32. Ruiz, H.; Cobo, P.; Jacobsen, F. Optimization of multiple-layer microperforated panels by simulated annealing. *Appl. Acoust.* **2011**, *72*, 772–776. [CrossRef]
33. Li, D.; Chang, D.; Liu, B. Enhancing the low frequency sound absorption of a perforated panel by parallel-arranged extended tubes. *Appl. Acoust.* **2016**, *102*, 126–132. [CrossRef]
34. Li, X.; Wu, Q.; Kang, L.; Liu, B. Design of Multiple Parallel-Arranged Perforated Panel Absorbers for Low Frequency Sound Absorption. *Materials* **2019**, *12*, 2099. [CrossRef]

materials MDPI

Article

Variability and Performance Study of the Sound Absorption of Used Cigarette Butts

Valentín Gómez Escobar [1,*], Guillermo Rey Gozalo [1,2] and Carlos J. Pérez [3]

1 Departamento de Física Aplicada, Escuela Politécnica, Universidad de Extremadura, Avda. de la Universidad s/n, 10003 Cáceres, Spain
2 Facultad de Ciencias de la Salud, Universidad Autónoma de Chile, 5 Poniente 1670, 3460000 Talca, Región del Maule, Chile
3 Departamento de Matemáticas, Facultad de Veterinaria, Universidad de Extremadura, Avda. de la Universidad s/n, 10003 Cáceres, Spain
* Correspondence: valentin@unex.es; Tel.: +34-927-257596; Fax: +34-927-257203

Received: 3 July 2019; Accepted: 12 August 2019; Published: 14 August 2019

check for updates

Abstract: There has been increasing interest in new sustainable materials that can be used as construction materials. Among them, sound-absorbing materials have an important role in both acoustical room conditioning and in room insulation. As a proposal for recycling, one of the most common residues in the world, cigarette butts, is studied. Samples were prepared with used cigarette butts as acoustical absorbent materials. Several samples were prepared and grouped by similarity. Variability analyses of the samples prepared in each group were performed. Moreover, the analysis of some possible influences on absorption properties, such as the length of butts, presence of burnt regions, presence of wrapping paper, etc., were analyzed. The results show the potentiality of this residue to be used as an acoustical absorbent since the absorption coefficients found are greater than 0.8 for frequencies over 2000 Hz. The observed variability in the study group and samples can be considered low, as it was below 2% for the major part of frequencies. Influences on the absorption coefficient, for both the length and status of the butts, were statistically confirmed.

Keywords: sound absorber; cigarette butts; sustainable material; recycling; variability analysis

1. Introduction

Among the different construction materials, there has been special interest in sound absorption materials [1] that can help both in adapting room acoustics to a particular use and in achieving better insulation characteristics of different building partitions. The use of industrially manufactured mineral wools is widely extended for sound absorption in construction although, in recent years, some alternative materials have been proposed for acoustical absorption [2]. Thus, there are studies on the acoustical behavior of indoor or outdoor plants [3–5], materials made of natural fibers [6–10], sheep wool [11], cork wastes [12], and olive pruning wastes [13], to name but a few.

In recent decades, there have been increasing efforts to recycle all the waste materials produced by human activities. This recycling practice has also emerged in construction materials. Thus, as a more sustainable option, among the different proposed alternative materials for sound absorption, some are based on recycling materials [12–15]. These recycled materials have the additional value of contributing to the elimination of waste materials produced by human activities.

One of the major components of human debris is used cigarette butts. Cigarette filters were introduced (in the middle of the past 20th century) to reduce the incidence of diseases in smokers. However, this had a collateral effect, i.e., the fact that the cellulose diacetate (substance from which filters are mainly composed) is not biodegradable and has only partial photodegradation makes used cigarette butts an important environmental concern. It must be considered that, throughout the world, billons of cigarettes are consumed every year [16,17], and it is expected that the consumption will increase by more than 50% by 2025 due to the increase in world population and tobacco production [18]. The problem due to this residue is enlarged due to the unfortunately common habit of smokers to throw cigarette butts on the ground [19,20]. Thus, in studies on the debris composition of places of garbage collection, such as bins, containers, etc., but also in studies of different human environments (streets, beaches, etc.), butts are commonly the main component in number and even sometimes in weight [21–23]. The butts thrown on the ground are transported by rain or river water to coastal areas, increasing their environmental impact. This environmental impact of used cigarette butts is related both to their chemical composition [24] and to the difficult degradation of cellulose diacetate, as mentioned. Some of the 130 chemicals found in cigarette butts have been described. In addition, many more chemicals (40,000–100,000) have been identified in cigarette smoke, some of which could be retained in cigarette butts [25]. These chemicals can easily leach into water, making these waters toxic for some organisms [24,26–28]. Consequently, cigarette butts are a concern in terms of the environment and public health.

There are some recycling proposals concerning cigarette butts, such as using them as part of the composition of bricks [29] or supercapacitors [30], or using their washing waters as chemical inhibitors [31] or as insecticide [32], etc. A revision of different proposals for recycling used butts has recently been carried out [33]. However, these proposals would not be enough to recycle all the annual production of this major residue and, certainly, new proposals are welcome. In this sense, our research group is working on the applying used cigarette butts as acoustical absorbers.

Some preliminary studies have shown the potential of used cigarettes butts for the preparation of acoustic absorbers [34,35]. However, additional studies are needed to analyze the application of used cigarette butts for this purpose. For example, whether the variability of the status (used and/or burned), diameter, and length of the butts can influence their acoustic properties. Additionally, the effect of the paper covering the cigarette filters is a factor to consider. Moreover, the comparison between used and unsmoked butts can also be interesting. These considerations are dealt with in this paper.

Taking into account the abovementioned information, the influence of the status, length, and paper wrapping on the spectrum of sound absorption are analyzed in the present study. For this purpose, preliminary and careful determination of the characteristics of the cigarette butts was carried out. Moreover, an appropriate number of similar samples were prepared, and the analysis of the different measured absorption coefficients is enriched through statistical analysis. A relevant influence of these factors on the absorption properties of the samples prepared with butts would imply allow for selective collection and planning for these residues.

2. Materials and Methods

2.1. Instrumentation for Acoustic Absorption Determination

The determination of the sound absorption coefficient of different samples was carried out using an impedance tube, following the two-microphone transfer function method described in the ISO 10534-2 standard [36].

The measurements were conducted using an Impedance Tube Kit (Type 4206, Brüel & Kjaer, Nærum, Denmark), equipped with two one-quarter-inch condenser microphones (Type 4187). As the prepared samples could be considered non-consolidated, the tube was placed in a vertical position (see Figure 1). The signals were analyzed using a portable PULSE System (Brüel & Kjaer, Nærum, Denmark) with four input data channels (Type 3560-C). A sample holder, with a diameter of 29 mm, was used (with a validity in the frequency range of 500 Hz to 6.4 kHz).

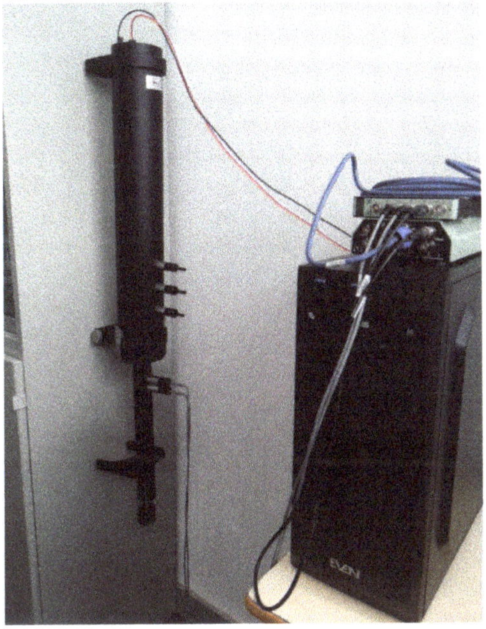

Figure 1. Impedance tube: experimental setup.

2.2. Preparation of Samples

Smoked cigarette butts were collected from ashtrays or the ground, and the remaining non-smoked tobacco was then manually separated. Only the cigarette butts were taken.

These remaining butts were nonhomogeneous. The length, diameter, presence of burnt regions, compression exerted by user's fingers, humidity, etc., could be different in each butt. In order to avoid part of the variability introduced by these characteristics, for this study, the cigarette butts were separated into equivalent groups prior to the preparation of the samples. The diameter and length of all the cigarette butts were measured before sample preparation. While there are some brands of cigarettes that use a narrower diameter, for the present study, a diameter of only around 7.5–8.0 mm was used, as it was the most common diameter among the available cigarette butts.

As the diameter of the butts can be considered very similar, the main variables to consider were the initial status of the butts (used, or unsmoked, and/or burnt) and their length. Three length groups were considered:

- Length 1 (L1): length = 15.13 ± 0.40 mm; diameter = 7.95 ± 0.07 mm.
- Length 2 (L2): length = 20.67 ± 0.22 mm; diameter = 7.64 ± 0.06 mm.
- Length 3 (L3): length = 26.61 ± 0.13 mm; diameter = 7.59 ± 0.04 mm.

The following four statuses were established:

- Condition 1 (C1): Smoked cigarette butts without burnt regions. Burnt regions were considered when some hard structure was formed in the cigarette butts due to the burning of a part of the filter.
- Condition 2 (C2): Smoked cigarette butts with burnt regions.
- Condition 3 (C3): Smoked cigarette butts without burnt regions and without the paper that wrapped the filter. The wrapping paper was removed carefully, avoiding loss of the fibrous material of the filter.
- Condition 4 (C4): Unsmoked cigarette butts.

Each sample was prepared by manually placing 10 cigarette butts in the sample holder of the impedance tube. This number of butts was selected to avoid high compression of butts in the samples. At least 10 cigarette butts samples from each group were prepared in order to have an adequate number of samples to allow for subsequent statistical analyses. The different groups of samples measured are presented in Table 1, indicating the number of samples prepared (each one with at least 10 different samples, as mentioned), density of samples, and length and diameters of the used cigarette for each group. Porosity of samples was determined considering the porosity of some individual butts (calculated by a helium pycnometer) and considering a spherical geometry for the cigarette butts of samples.

Table 1. Some properties (mean ± standard deviation) of the prepared samples. L = length; ϕ = diameter; ρ =density; por = porosity.

Length (L)	Cigarette Butts Status			
	C1 (Smoked without Burnt)	C2 (Smoked with Burnt)	C3 (Smoked without Burnt without Wrapping Paper)	C4 (Unsmoked with Wrapping Paper)
L1	20 samples L = 15.4 ± 0.1 mm ϕ = 8.01 ± 0.02 mm ρ = 140 ± 4 kg/m^3 por = 0.929 ± 0.003	–	–	14 samples L = 14.7 ± 0.1 mm ϕ = 7.88 ± 0.01 mm ρ = 97 ± 2 kg/m^3 por = 0.931 ± 0.001
L2	20 samples L = 20.8 ± 0.2 mm ϕ = 7.65 ± 0.07 mm ρ = 118 ± 7 kg/m^3 por = 0.935 ± 0.008	20 samples L = 20.5 ± 0.2 mm ϕ = 7.63 ± 0.04 mm ρ = 123 ± 5 kg/m^3 por = 0.936 ± 0.005	10 samples L = 20.9 ± 0.2 mm ϕ = 7.63 ± 0.08 mm ρ = 79 ± 3 kg/m^3 por = 0.936 ± 0.010	–
L3	20 samples L = 26.6 ± 0.2 mm ϕ = 7.59 ± 0.04 mm ρ = 114 ± 2 kg/m^3 por = 0.936 ± 0.005	–	–	–

Some of the prepared samples are shown in Figure 2. As can be seen, the samples present a particular appearance, as they are nonhomogeneous. Indeed, three different porosities can be distinguished: Firstly, the porosity due to the presence of voids among the different butts and among butts and the impedance tube wall (we can call it external porosity); secondly, the porosity peculiar to the butts (voids between the fibers of the butts) (filter porosity); and, finally, the porosity that the fibers that compound the filter of the cigarette butt could have (fiber porosity). This triple porosity gives the samples a complicated structure.

(a) (b) (c) (d)

Figure 2. Example pictures of some of the prepared samples: (**a**) Sample from C1 cigarette butts; (**b**) Sample from C2 cigarette butts; (**c**) Sample from C3 cigarette butts; (**d**) Sample from C4 cigarette butts.

2.3. Statistical Analysis

A variability analysis was performed based on the curves of the absorption coefficients and their Pearson's coefficient of variation (CV). Descriptive statistics, such as the mean, standard deviation, and minimum and maximum values have been considered. Additionally, 95% confidence intervals for the mean value of the absorption coefficient are presented.

The coefficient of variation (CV) is defined as the relationship between the standard deviation of the sample and the average value. It measures the relative dispersion of the data (to what extent the data is close or far from its average). As this coefficient is dimensionless, it can be used to compare different datasets regardless of the units of measure.

An analysis of variance for the functional data (functional ANOVA) has been considered to report statistically significant differences between the curves of different groups of samples in the full frequency spectrum [37]. Data were transformed to 1/3 of the octave bands and one-way ANOVA and a *t*-test (with Bonferroni correction) have been applied to determine if there were statistically significant differences among the groups in those frequency bands.

Statistical analyses were performed using R software, release 3.6.0 [38]. Package fdANOVA [39] was used for the functional ANOVA. Two-sided $p < 0.05$ were considered statistically significant.

3. Results and Discussion

Firstly, a variability analysis is performed, then combinations of samples are compared and, finally, the one-third octave bands' absorption coefficients were analyzed.

3.1. Variability Analysis

As can be seen in Table 1, several samples were prepared for each of the six groups. As mentioned, the cigarette butts used for the preparation of each group of samples have some similar characteristics (e.g., similar status, length, and diameter). Nevertheless, as the used butts can have a different precedence (brand, weather exposure, etc.) or a different compression exerted by user's fingers, humidity, etc., it is important to study the observed variability among the different samples inside each combination.

In a first step, all the absorption coefficients of the six different combinations of samples shown in Table 1 were represented (see Figure 3).

As can be seen in Figure 3, samples from each group do not present great variability. Despite the generally low variability, it was not the same for all the frequencies, as in some parts of the spectrum, the absorption coefficients of the different samples are more similar in value than in other parts. All measurements present a first maximum in the absorption coefficient, and the variability was higher under this absorption maximum. An increase in the variability can also be observed at higher frequencies (over 4500 Hz). Comparing the different groups, the variability of the graphs of samples with burnt butts (Figure 3d) and that of samples without paper (Figure 3e) was higher than the variability of the rest of the graphs.

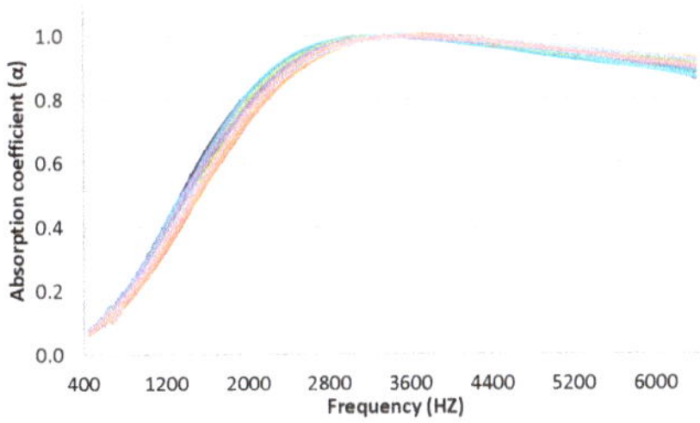

(a)

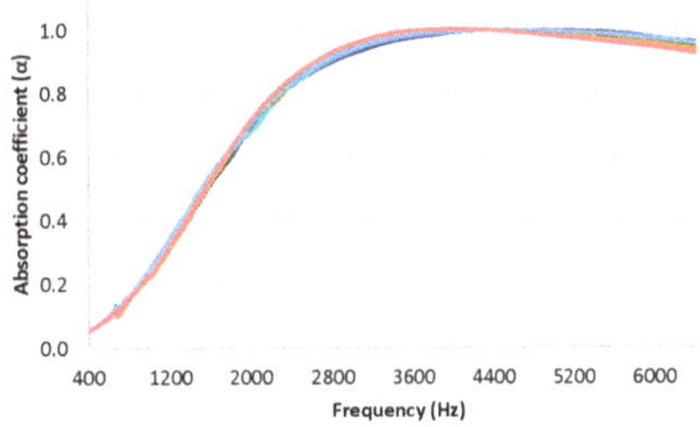

(b)

Figure 3. *Cont.*

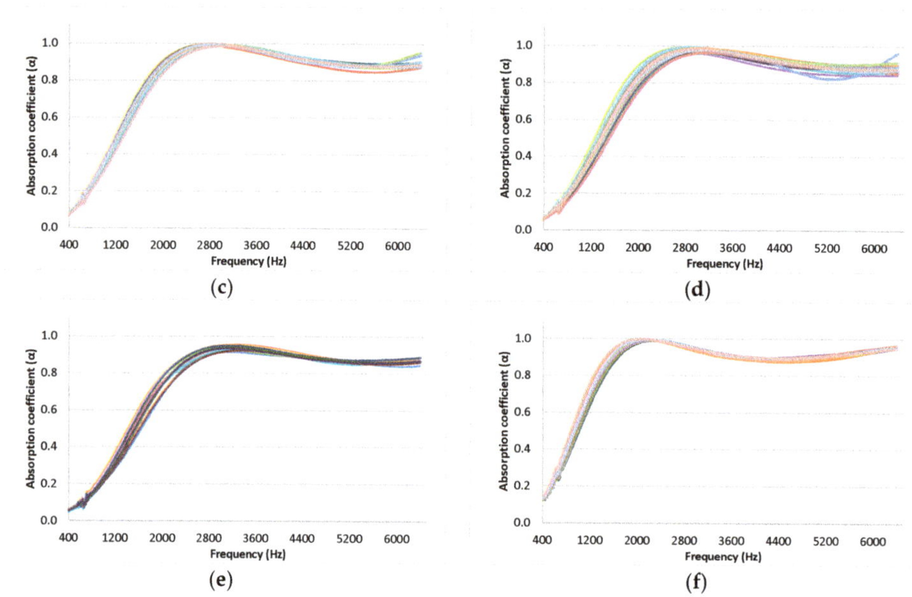

Figure 3. Absorption coefficient of the measured samples: (**a**) Samples from Group L1–C1; (**b**) Samples from Group L1–C4; (**c**) Samples from Group L2–C1; (**d**) Samples from Group L2–C2; (**e**) Samples from Group L2–C3; (**f**) Samples from Group L3–C1.

The results shown in Figure 3 were then analyzed through an inferential analysis. For this purpose, 95% confidence intervals for the mean value of the absorption coefficient were calculated. The confidence intervals were very close to the mean, as shown in Figure 4. Therefore, these results confirm the low variability previously indicated in Figure 3.

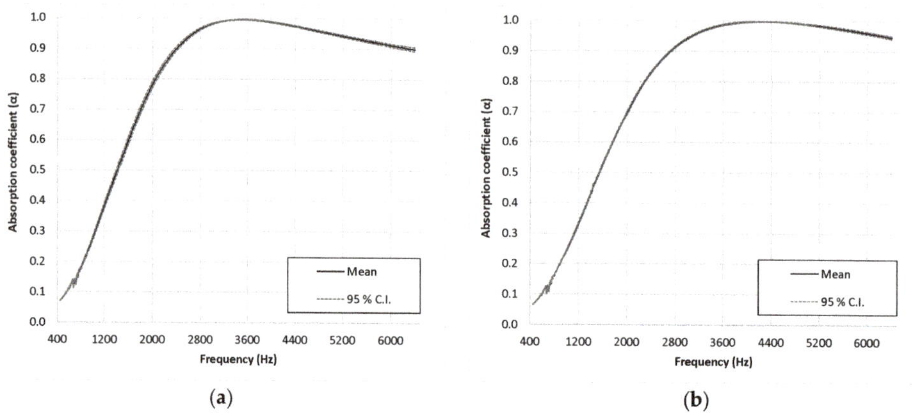

Figure 4. *Cont.*

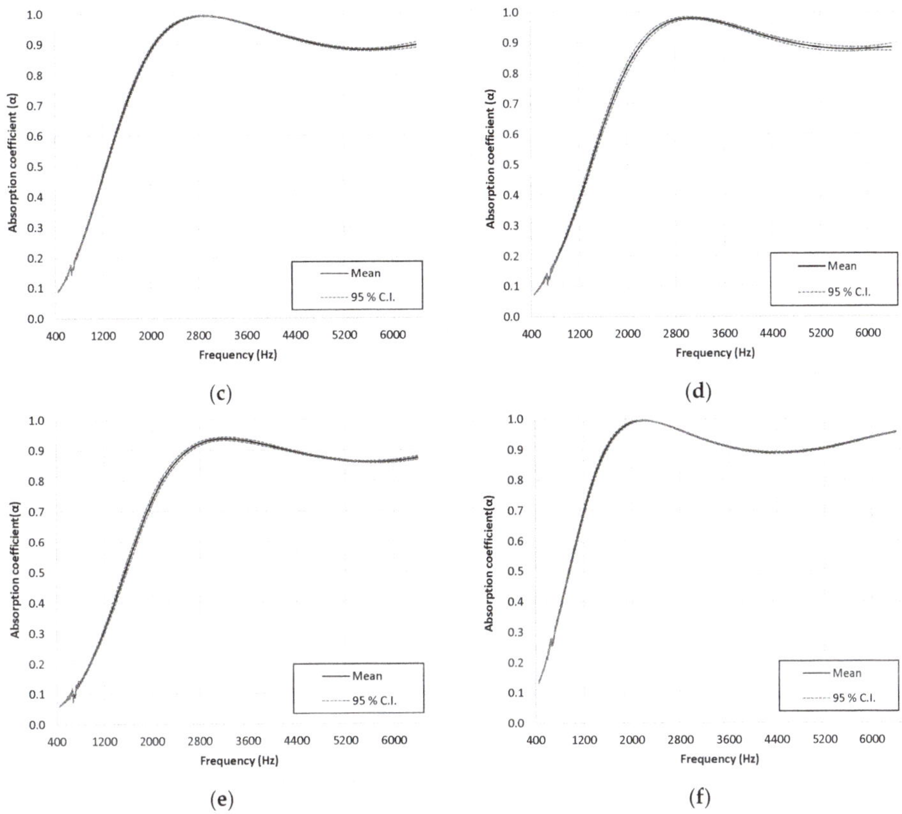

Figure 4. Mean value curves of the absorption coefficient, with 95% confidence interval curves: (**a**) Group L1–C1; (**b**) Group L1–C4; (**c**) Group L2–C1; (**d**) Group L2–C2; (**e**) Group L2–C3; (**f**) Group L3–C1.

In addition to analyzing the absolute variability, the relative variability was evaluated using the coefficient of variation (CV). The CV obtained for the different frequencies is shown in Figure 5 and a descriptive analysis of the obtained values (mean, standard deviation, maximum, and minimum) is shown in Table 2.

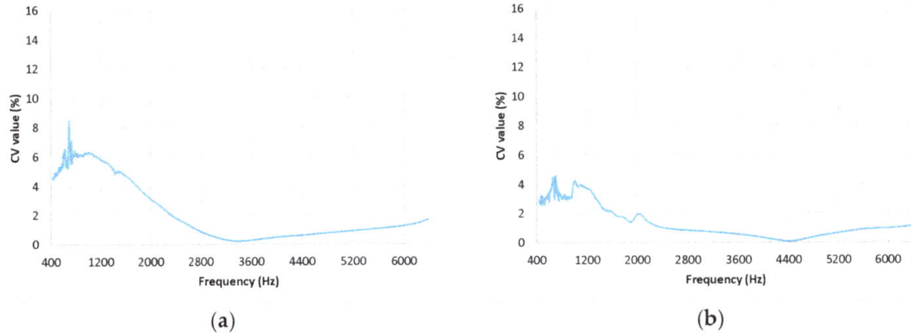

Figure 5. *Cont.*

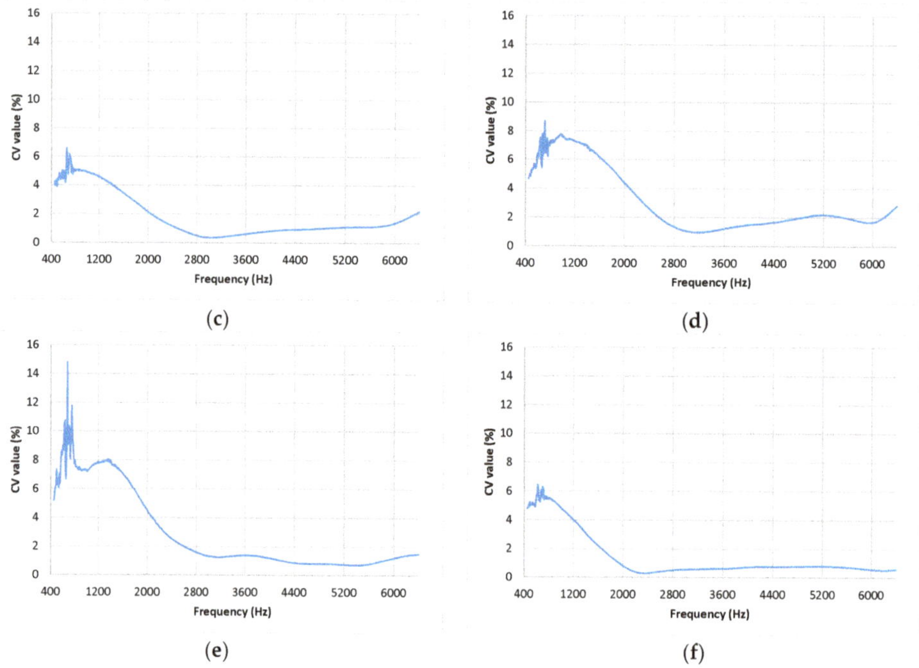

Figure 5. Coefficient of variation curves of the coefficient of absorption for the different groups of analyzed samples: (**a**) Group L1–C1; (**b**) Group L1–C4; (**c**) Group L2–C1; (**d**) Group L2–C2; (**e**) Group L2–C3; (**f**) Group L3–C1.

Table 2. Average, standard deviation, and maximum and minimum of the CV values in a range from 504 to 6400 Hz for each group.

Group	Average	Standard Deviation	Minimum	Maximum
L1–C1	2.08	2.01	0.24	8.48
L1–C4	1.28	1.08	0.10	4.63
L2–C1	1.80	1.49	0.35	6.61
L2–C2	3.03	2.17	0.94	8.66
L2–C3	2.90	2.77	0.71	14.79
L3–C1	1.37	1.53	0.26	6.41

As in the previous analyses, in Figure 5 and Table 2, it can be observed that all groups of samples have a low coefficient of variation. The coefficients of variability were less than 10% in all cases, except in some frequencies of group L2–C3, and in all cases, the averaged CV was lower than 3.03%. Comparing the different combinations of samples, the highest variability was found in the samples of groups L2–C2 and L2–C3. Group L2–C2 corresponds to burnt butts, and the fact that the size and position of the burnt zones were not equal in the different butts can explain this higher variability among the samples prepared with burnt cigarette butts. Group L2–C3 corresponds to the samples prepared with butts without the wrapping paper; when the paper is removed, the cigarette sometimes butt lost its shape, and this can be the reason for the increase in the variability among the samples of this group. With respect to frequencies, the range from 1000 to 2000 Hz is the most variable. At higher frequencies, a decrease in variability is first observed, then a slight new increase of variability is observed from 4400 Hz. Nevertheless, in general, the coefficient of variation is less than 2% from 2800 Hz onwards.

3.2. Comparison between Samples

In order to analyze whether the presence of burnt regions, the presence of wrapping paper, or the length in the butts of the samples have a significant influence on the absorption properties of the samples, we carried out a comparison of the groups of samples that could be considered similar in some way (located in the same row or in the same column in Table 1).

To analyze the influence of the length of butts in relation to the acoustical behavior of the samples, the average absorption coefficient for the three combinations of the samples belonging to status 1 (the mean values of the absorption coefficient measured for these samples are shown in Figure 6) were compared.

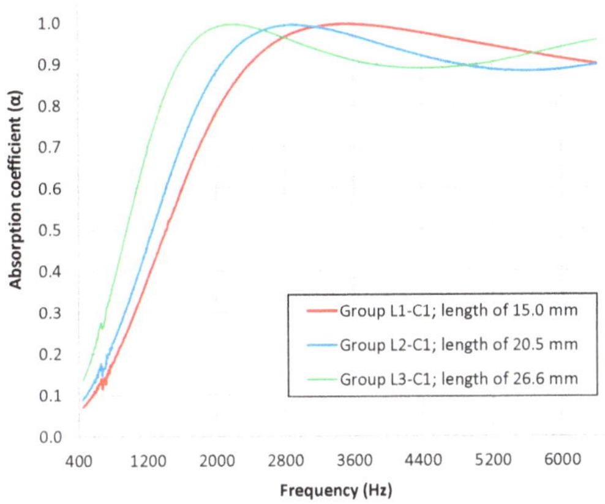

Figure 6. Mean value curves of the absorption coefficient for the three combinations of the samples of status 1.

It is known that the absorption of a porous material is due to the energy losses of the sound waves as moving air particles, induced by the wave, interact with the motionless particles near the skeleton or fibers of the porous material. Thus, the absorption is at its maximum when the maximum particle speed is located inside the absorbent material. As in the disposition of the measured samples, the porous material is faced with the rigid termination of the impedance tube, and near this rigid termination, there is a minimum particle speed with the maximum being located at a certain distance (λ/4) from this termination. Thus, when the thickness of the sample increases, the maximum absorption of the material will be shifted to lower frequencies. This shift was observed in the measured samples, as can be seen in Figure 6. Comparing the absorption coefficient curves, statistically significant differences were found ($p < 0.001$) using functional ANOVA for the three groups of samples. The shift in the maximum absorption frequency with the thickness of butt samples was studied in a previous work [34] showing that the relationship observed for butt samples was different to the observed for other sample (glass wool). The shift observed in the samples or the present research was congruent with the observed in a previous work [34] as can be observed in Figure 7, where results of both studies are compared.

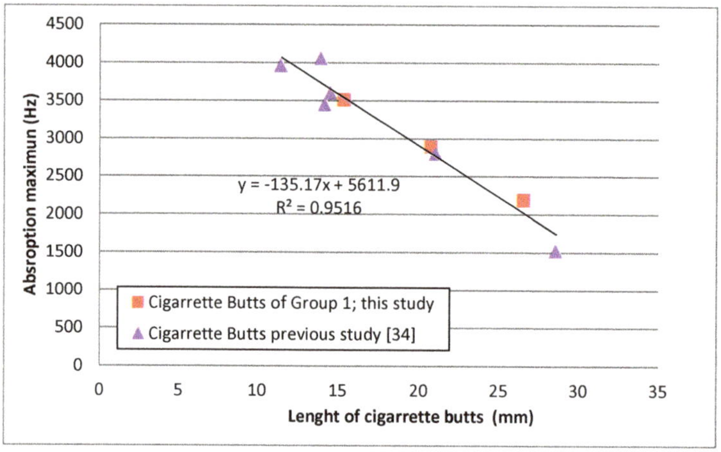

Figure 7. Variation of the frequency of maximum absorption with the length of cigarette butts of the samples.

To analyze the influence of the presence of burnt regions and the effect of removing the wrapping paper on the absorption coefficient, the average absorption coefficient of these two combinations of samples (groups L2–C2 and L2–C3) were compared with the group of used butts with wrapped paper and without burnt regions (group L2–C1) (Figure 8). All these samples were composed of used butts with a similar length (about 20.5 mm).

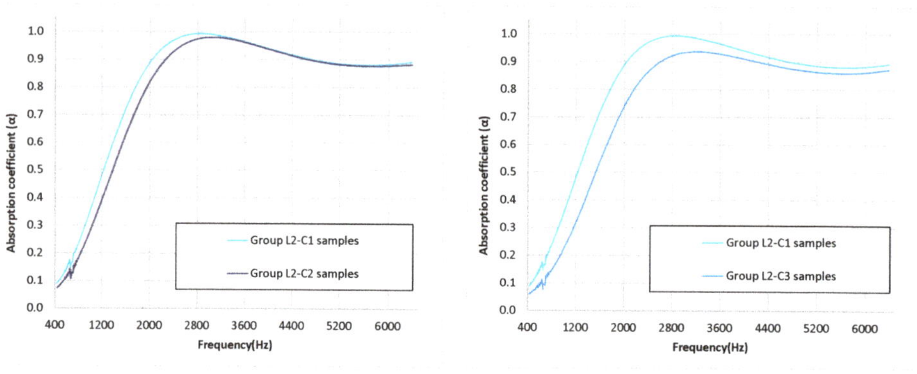

Figure 8. Mean value of the absorption coefficient for the three combinations of samples of lengths of around 20.5 mm. (**a**) Samples of used butts with paper and without burnt regions (Group L2–C1) are compared with samples of burnt butts (Group L2–C2); (**b**) Samples of used butts with paper and without burnt regions (Group L2–C1) are compared with samples composed of butts without paper and without burnt regions (Group L2–C3).

As can be seen in Figure 8a, the maximum of absorption occurs at about the same frequency (around 3000 Hz) for both samples with and without burnt butts, and a similar absorption coefficient is observed for higher frequencies. Nevertheless, the absorption at frequencies under this maximum value were clearly lower than in the case of samples composed of used butts without burnt regions. The slight decrease in the absorption coefficient in the maximum value and the shift of this maximum to higher frequencies could be explained by considering that burnt regions represent a decrease in the number of fibers (they have collapsed in the significantly burnt regions) and, thus, the absorption efficiency of the material is reduced. This reduction of the absorption efficiency is not appreciable at higher frequencies, probably due to the fact that, at those frequencies, more than one maximum particle speed is inside the material. Despite the similar behavior at high frequencies, as mentioned, the functional ANOVA showed statistically significant differences ($p < 0.001$) between both groups (groups L2–C1 and L2–C2).

Regarding the effect of the wrapping paper, in Figure 8b, it can be seen that samples without this wrapping paper present both a lower absorption coefficient in all spectra, although the differences are lower with the frequency of the maximum absorption value. Additionally, a shift of the value of the frequency of the maximum of absorption to higher frequency values can be observed. The first effect (a decrease in the absorption efficiency) could be explained by the own absorption of the wrapping paper. It is important to note that the weight in the samples decreased by 34.1% ± 2.4% when the wrapping paper was removed. The second effect (a shift in the frequency of the maximum absorption to higher frequencies) would be consistent with the described decrease in the thickness of the sample, but the length of the butt of the samples was the same in both groups of samples. This shift can be explained, in this case, by the variation of the density of the absorber in the samples, which also produces a shift in the frequency of the maximum absorption, as previously described [34]. Due to important differences between both groups of samples, statistical differences were expected; thus, functional ANOVA also showed statistically significant differences ($p < 0.001$) between both sample groups (groups L2–C1 and L2–C3).

Finally, to evaluate the effect of smoking processes on the absorption properties of butts, samples composed of unsmoked butts (group L1–C4) were compared with samples composed of used butts with a similar length (group L1–C1) (both with butts of a length of approximately 15 mm) (see Figure 9). As can be seen in this figure, samples composed of unsmoked butts present a higher absorption with the frequency of the maximum absorption value than samples composed of used butts. Moreover, there is a shift of this maximum of absorption to higher frequencies in the case of unsmoked butt samples. As the unsmoked butt length is slightly lower than the length of the used ones with which they are being compared, the difference may also be explained by this difference in length, as shown in Figure 6. In this case, functional ANOVA also showed statistically significant differences ($p < 0.001$) between both sample groups (groups L1–C1 and L1–C4).

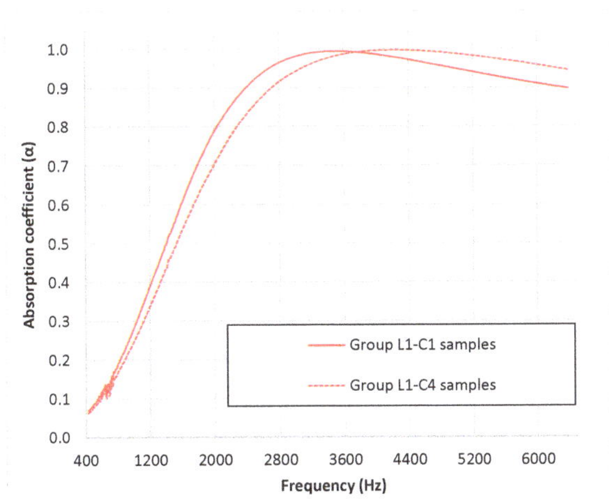

Figure 9. Mean value of the absorption coefficient for the two combinations of samples of a length of around 15.0 mm.

3.3. Analysis of the One-Third Octave Bands' Absorption Coefficients

While functional ANOVA indicated that the overall absorption coefficients presented statistically significant differences among all the compared groups, in some frequency ranges of the graphs (i.e., for the high frequencies in Figure 8a), the observed curves were similar. For this reason, the initial coefficients of the absorption data were transformed into one-third octave bands, and then one-way ANOVA and a *t*-test (with Bonferroni correction) were used to compare the one-third octave means in the different groups of samples. The obtained *p*-values are shown in Table 3.

Table 3. Obtained *p*-values for the *t*-test, with Bonferroni correction, for the different comparisons. Differences that are not statistically significant are presented in bold.

	Obtained *p*-values					
1/3 Octave Band (Hz)	**L1–C1 vs. L2–C1**	**L2–C1 vs. L3–C1**	**L1–C1 vs. L3–C1**	**L2–C1 vs. L2–C2**	**L2–C1 vs. L2–C3**	**L1–C1 vs. L1–C4**
500	<0.001	<0.001	<0.001	<0.001	<0.001	<0.001
630	<0.001	<0.001	<0.001	<0.001	<0.001	<0.001
800	<0.001	<0.001	<0.001	<0.001	<0.001	<0.001
1000	<0.001	<0.001	<0.001	<0.001	<0.001	<0.001
1250	<0.001	<0.001	<0.001	<0.001	<0.001	<0.001
1600	<0.001	<0.001	<0.001	<0.001	<0.001	<0.001
2000	<0.001	<0.001	<0.001	<0.001	<0.001	<0.001
2500	<0.001	**0.26**	<0.001	<0.001	<0.001	<0.001
3150	**0.53**	<0.001	<0.001	<0.001	<0.001	<0.001
4000	<0.001	<0.001	<0.001	**0.30**	<0.001	<0.001
5000	<0.001	<0.01	<0.001	**0.49**	<0.001	<0.001
6400	<0.001	<0.001	<0.001	**0.08**	<0.001	<0.001

As can be seen, there is only one one-third octave band frequency in which differences were not statistically significant between Group L2–C1 (20.5 mm) and Group L1–C1 (15.0 mm) and, moreover, only one between Group L2–C1 (20.5 mm) and Group L3–C1 (26.6 mm). These two one-third octave band frequencies correspond to the overlap among the absorption coefficient data near the maximum value, which can be observed in Figure 6. With respect to Figure 7, the Group L2–C1 and Group L2–C2 results overlap at higher frequencies (4000, 5000, and 6400 Hz one-third octave bands) and, so, for these one-third octave band frequencies, no statistically significant differences were observed. For the rest of the pairwise comparisons, all the one-third octave bands presented statistically significant differences.

4. Conclusions

The studies, carried out with more than 100 samples, showed that samples prepared with used cigarette butts presented a high absorption coefficient. Thus, the obtained absorption coefficient values showed that the absorption capacity of the different measured samples were quite high, with absorption coefficients higher than 0.8 for frequencies over 2000 Hz.

According to this study, the absorption efficiency depends on the presence of burnt regions, the removal of the paper that wrapped the filter, and the length of the butts used for sample preparation. This effect was lower when the frequency was higher than the maximum of the absorption frequency.

Shifts in the frequency corresponding to the maximum value of the coefficient absorption were observed. These shifts could be explained by the different lengths of cigarette butts used in the sample preparation and the different densities of absorbing materials. This is consistent with the study presented in [34].

While some characteristics of the used butts could be different between samples (for instance, the humidity, degree of splashing, brand of the butts, etc.), the variability found in the measured samples can be considered low. Despite the low observed values of variability, it was observed that the presence of burnt regions in the used butt and the removal of the wrapped paper increased the variability of the samples.

5. Limitations of the Study and Further Studies

The results of this study, despite their potential, present some limitations. Firstly, the obtained absorption coefficients are valid only for perpendicular incidence. For real applications, a random incidence absorption coefficient would be desirable. Secondly, the absorption coefficients must be accompanied by other measured parameters such as flow resistivity, tortuosity, sample porosity, etc., in order to understand the acoustical behavior of samples.

With respect to the sample preparation, the chosen distribution of butts will probably not be adequate for a higher number of samples and other possibly more homogenous samples would be desirable.

To complete the analysis of the potential to use this recycled product in sound absorption applications in the construction industry, further studies are necessary which examine samples prepared with other configurations and samples prepared after cleaning the used butts prior to sampling preparation. Moreover, better knowledge of the properties of the samples (properties such as flow resistance, porosity, or tortuosity) would be advisable. Finally, comparisons with other conventional or sustainable materials would also be necessary.

Author Contributions: V.G.E. designed and performed the acoustical measurements. C.J.P. and G.R.G. designed and performed the statistical analysis of the data. All the authors contributed in a similar way in the original drafting preparation and in the review and editing of the final document.

Funding: This research was partially funded by Altadis (Imperial Tobacco), by the European Regional Development Fund (ERDF), by Junta de Extremadura, Consejería de Economía, e Infraestructura (Project IB18033), by Agencia Estatal de Investigación, Spain (Project MTM2017-86875-C3-2-R) and by Junta de Extremadura, Spain (GR18108 and GR18107). Guillermo Rey Gozalo was supported by the Juan de la Cierva–Incorporación contract from the Spanish Ministry of Economy, Industry and Competitiveness (IJCI-2016-28923).

Acknowledgments: The authors would like to thank the researcher Celia Moreno González for her help in the sample preparation and measurement.

Conflicts of Interest: The authors declare no conflict of interest.

References

1. Attenborough, K.; Ver, I.L. Sound-absorbing materials and sound absorbers. In *Noise and Vibration Control Engineering*, 2nd ed.; Wiley: New York, NY, USA, 2006; Chapter 8.
2. Asdrubali, F.; Schiavoni, S.; Horoshenkov, K.V.A. Review of Sustainable Materials for Acoustic Applications. *Build. Acoust.* **2012**, *19*, 283–312. [CrossRef]
3. Attenborough, K.; Bashir, I.; Taherzadeh, S.; Hill, T. Models for acoustical properties of green roof materials. In Proceedings of the Forum Acusticum 2011, Aalborg, Denmark, 27 June–1 July 2011; pp. 863–868.
4. Iannace, G.; Trematerra, A.; Trematerra, P. Acoustic correction using green material in classrooms located in historical buildings. *Acoust. Aust.* **2013**, *41*, 213–218.
5. D'Alessandro, F.; Asdrubali, F.; Mencarelli, N. Experimental evaluation and modelling of the sound absorption properties of plants for indoor acoustics applications. *Build. Environ.* **2015**, *94*, 913–923. [CrossRef]
6. Glé, P.; Gourdon, E.; Arnaud, L. Acoustical properties of materials made of vegetable particles with several scales of porosity. *Appl. Acoust.* **2011**, *72*, 249–259. [CrossRef]
7. Oldham, D.J.; Egan, C.A.; Cookson, R.D. Sustainable acoustic absorbers from the biomass. *Appl. Acoust.* **2011**, *72*, 350–363. [CrossRef]
8. Berardi, U.; Iannace, G. Acoustical characterization of natural fibers for sound absorption applications. *Build. Environ.* **2015**, *94*, 840–852. [CrossRef]
9. Mamtaz, H.; Fouladi, M.H.; Al-Altabi, M.; Namasivayam, N. Acoustic absorption of natural fiber composites. *J. Eng.* **2016**, *2016*. [CrossRef]
10. Berardi, U.; Iannace, G. Predicting the sound absorption of natural materials: Best-fit inverse laws for the acoustic impedance and the propagation constant. *Appl. Acoust.* **2017**, *115*, 131–138. [CrossRef]

11. Del Rey, R.; Uris, A.; Alba, J.; Candelas, P. Characterization of sheep wool as a sustainable material for acoustic applications. *Materials* **2017**, *10*, 1277. [CrossRef]
12. Maderuelo-Sanz, R.; Barrigón Morillas, J.M.; Gómez Escobar, V. Acoustical performance of loose cork granulates. *Eur. J. Wood Wood Prod.* **2014**, *72*, 321–330. [CrossRef]
13. Martellotta, F.; Cannavale, A.; De Matteis, V.; Ayr, U. Sustainable sound absorbers obtained from olive pruning wastes and chitosan binder. *Appl. Acoust.* **2018**, *141*, 71–78. [CrossRef]
14. Maderuelo-Sanz, R.; Barrigón Morillas, J.M.; Martín Castizo, M.; Gómez Escobar, V.; Rey Gozalo, G. Acoustical performance of porous absorber made from recycled rubber and polyurethane resin. *Lat. Am. J. Solids Struct.* **2013**, *10*, 585–600. [CrossRef]
15. Khan, A.; Mohamed, M.; Al Halo, N.; Benkreira, H. Acoustical properties of novel sound absorbers made from recycled granulates. *Appl. Acoust.* **2017**, *127*, 80–88. [CrossRef]
16. Euromonitor International. Passport: Global Market Information Database. 2016. Available online: http://www.tobaccoatlas.org/ (accessed on 11 June 2019).
17. Novotny, T.E.; Zhao, F. Consumption and production waste: Another externality of tobacco use. *Tob. Control* **1999**, *8*, 75–80. [CrossRef]
18. Mackay, J.; Eriksen, M.; Shafey, O. *The Tobacco Atlas*, 2nd ed.; American Cancer Society: Atlanta, GA, USA, 2000; pp. 32–33.
19. Novotny, T.E.; Slaugther, E. Tobacco Product Waste: An Environmental Approach to Reduce Tobacco Consumption. *Curr. Environ. Health Rep.* **2014**, *1*, 208–216. [CrossRef]
20. Rath, J.M.; Rubenstein, R.A.; Curry, L.E.; Shank, S.E.; Cartwright, J.C. Cigarette Litter: Smokers' Attitudes and Behaviors. *Int. J. Environ. Res. Public Health* **2012**, *9*, 2189–2203. [CrossRef]
21. Ariza, E.; Jiménez, J.A.; Sardá, R. Seasonal evolution of beach waste and litter during the bathing season on the Catalan coast. *Waste Manag.* **2008**, *28*, 2604–2613. [CrossRef]
22. Martínez-Ribes, L.; Basterretxea, G.; Palmer, M.; Tintoré, J. Origin and abundance of beach debris in the Balearic Islands. *Sci. Mar.* **2007**, *71*, 305–314. [CrossRef]
23. Moriwaki, H.; Kitajima, S.; Katahira, K. Waste on the roadside, 'poi-sute' waste: Its distribution and elution potential of pollutants into environment. *Waste Manag.* **2009**, *29*, 1192–1197. [CrossRef]
24. Slaughter, E.; Gersberg, R.M.; Watanabe, K.; Rudolph, J.; Stransky, C.; Novotny, T.E. Toxicity of cigarette butts, and their chemical components, to marine and freshwater fish. *Tob. Control* **2011**, *20*, i25–i29. [CrossRef]
25. Gong, M.; Khurshid, S.; Poppendiecky, D. What's in a butt? Environmental contamination from airborne cigarette butt emissions. *Integr. Environ. Assess. Manag.* **2017**, *13*, 549–551. [CrossRef]
26. Register, K.M. Cigarette Butts as Litter- Toxic as Well as Ugly. *Bull. Am. Met. Soc.* **2000**, *25*, 23–29.
27. Micevska, T.; Warne, M.S.T.J.; Pablo, F.; Patra, R. Variation in, and Causes of, Toxicity of Cigarette Butts to a Cladoceran and Microtox. *Arch. Environ. Contam. Toxicol.* **2006**, *50*, 205–212. [CrossRef]
28. Lee, W.; Lee, C.C. Development toxicity of cigarette butts-An underdeveloped issue. *Ecotoxicol. Environ. Saf.* **2015**, *113*, 362–368. [CrossRef]
29. Mohajerani, A.; Kadir, A.A.; Larobina, L. A practical proposal for solving world's cigarette butt problem: Recycling in fired clay bricks. *Waste Manag.* **2016**, *52*, 228–244. [CrossRef]
30. Lee, M.; Kim, G.-P.; Don Song, H.; Park, S.; Yi, J. Preparation of energy storage material derived from a used cigarette filter for a supercapacitor electrode. *Nanotechnology* **2014**, *25*, 345601. [CrossRef]
31. Zhao, J.; Zhang, N.; Qu, C.; Wu, X.; Zhang, J.; Zhang, X. Cigarette butts and their application in corrosion inhibition for N80 Steel at 90 °C in a hydrochloric acid solution. *Ind. Eng. Chem. Res.* **2010**, *49*, 3986–3991. [CrossRef]
32. Murugan, K.; Suresh, U.; Panneerselvam, C.; Rajaganesh, R.; Roni, M.; Aziz, A.T.; Hwang, J.-S.; Sathishkumar, K.; Rajasekar, A.; Kumar, S.; et al. Managing wastes as green resources: Cigarette butt-synthesized pesticides are highly toxic to malaria vectors with little impact on predatory copepods. *Environ. Sci. Pollut. Res.* **2018**, *25*, 10456–10470. [CrossRef]
33. Torkashvand, J.; Farzadkia, M. A systematic review on cigarette butt management as a hazardous waste and prevalent litter: Control and recycling. *Environ. Sci. Pollut. Res.* **2019**, *26*, 11618–11630. [CrossRef]
34. Gómez Escobar, V.; Maderuelo-Sanz, R. Acoustical performance of samples prepared with cigarette butts. *Appl. Acoust.* **2017**, *125*, 166–172. [CrossRef]
35. Maderuelo-Sanz, R.; Gómez Escobar, V.; Meneses-Rodríguez, J.M. Potential use of cigarette filters as sound porous absorber. *Appl. Acoust.* **2018**, *129*, 86–91. [CrossRef]

36. ISO 10534-2. *Acoustics: Determination of Sound Absorption Coefficient and Impedance in Impedances Tubes. Part 2: Transfer-Function Method*; ISO: Geneva, Switzerland, 1998.
37. Cuevas, A.; Febrero, M.; Fraiman, R. An ANOVA test for functional data. *Comput. Stat. Data Anal.* **2004**, *47*, 111–122. [CrossRef]
38. R Core Team. *R: A Language and Environment for Statistical Computing*; R Foundation for Statistical Computing: Vienna, Austria, 2019; Available online: https://www.R-project.org/ (accessed on 11 June 2019).
39. Gorecki, T.; Smaga, L. fdANOVA: Analysis of Variance for Univariate and Multivariate Functional Data. R Package Version 0.1.2. 2018. Available online: https://CRAN.R-project.org/package=fdANOVA (accessed on 11 June 2019).

materials

Article

Composite Eco-Friendly Sound Absorbing Materials Made of Recycled Textile Waste and Biopolymers †

Chiara Rubino [1], Marilés Bonet Aracil [2], Jaime Gisbert-Payá [2], Stefania Liuzzi [1], Pietro Stefanizzi [1], Manuel Zamorano Cantó [2] and Francesco Martellotta [1,*]

1 Dipartimento di Scienze dell'Ingegneria Civile e dell'Architettura, Politecnico di Bari, via Orabona 4, I-70125 Bari, Italy; chiara.rubino@poliba.it (C.R.); stefania.liuzzi@poliba.it (S.L.); pietro.stefanizzi@poliba.it (P.S.)

2 Grupo de Investigación en la Industria Textil (GIITEX), Departamento de Ingeniería Textil y Papelera, Universitat Politècnica de València, 46022 Alcoy, Alicante, Spain; maboar@txp.upv.es (M.B.A.); jaigispa@txp.upv.es (J.G.-P.); mzamoran@upvnet.upv.es (M.Z.C.)

* Correspondence: francesco.martellotta@poliba.it; Tel.: +39-080-596-3631

† Portions of this paper have been presented in the conference paper "Preliminary investigation on the acoustic properties of absorbers made of recycled textile fibers", by Rubino C., Bonet-Aracil M., Liuzzi S., Martellotta F., in Proceedings of the 23rd International Congress on Acoustics, integrating 4th EAA Euroregio 2019, 9–13 September 2019, Aachen, Germany.

Received: 6 November 2019; Accepted: 29 November 2019; Published: 3 December 2019

Abstract: In recent years, the interest in reusing recycled fibers as building materials has been growing as a consequence of their ability to reduce the production of waste and the use of virgin resources, taking advantage of the potential that fibrous materials may offer to improve thermal and acoustic comfort. Composite panels, made of 100% wool waste fibers and bound by means of either a chitosan solution and a gum Arabic solution, were tested and characterized in terms of acoustic and non-acoustic properties. Samples with a 5 cm thickness and different density values were made to investigate the influence of flow resistivity on the final performance. Experimental results demonstrated that the samples had thermal conductivity ranging between 0.049 and 0.060 W/(m K), well comparable to conventional building materials. Similarly, acoustic results were very promising, showing absorption coefficients that, for the given thickness, were generally higher than 0.5 from 500 Hz on, and higher than 0.9 from 1 kHz on. Finally, the effects of the non-acoustic properties and of the air gap behind the samples on the acoustic behavior were also analyzed, proving that the agreement with absorption values predicted by empirical models was also very good.

Keywords: textile waste; biopolymers; sound absorption; sustainable materials; circular economy

1. Introduction

Materials used in the construction industry for thermal insulation and noise control are mainly inorganic and synthetic composites, i.e., glass wool, stone wool, and polystyrene. These materials, despite having low thermal conductivity values and high sound absorption coefficients, cause significant environmental impacts during their production processes [1] and may affect human health depending on fiber dose, dimension, and durability [2]. Conversely, one of the greatest challenges for future buildings is to guarantee low energy use, preserving indoor thermal and acoustic comfort by using bio-based components able to ensure a healthy and sustainable environment. Choosing non-toxic, environment friendly, and recyclable building composites implies a growing attention in testing natural (vegetal/animal) or waste fibers as an alternative to mineral and synthetic ones [1].

Many researchers have been carrying out studies about innovative "green" building materials. The difficult disposal of agro-residues has stimulated the interest of the research community in the possible use of agricultural waste products as fibrous matrix of bio-based composites [3]. Furthermore, the wide availability of many "local" natural materials encouraged several researchers to study their sound absorption and thermal insulation properties. Thus, olive tree leaves [4], flax, hemp [5], cotton stalk [6], straw bales [7], kenaf fibers [8], coir fibers [9], rice straw [10], and others [11] have been considered.

The disposal issue concerns agricultural by-products, as well as the apparel industry, which is responsible of an enormous volume of wastes resulting from a fast-changing fashion culture based on designing garments characterized by an intrinsic obsolescence. A textile fabric can provoke pollution from the earliest manufacturing process (pre-consumer waste) to the end of its useful life (post-consumer waste) [12]. Thus, the search for a proper way to reuse textile fibers is stimulating several attempts to use them in the construction industry, especially in the form of sustainable panels [13,14] or mats [15]. The reuse of waste or by-products as new raw materials for innovative and sustainable building components has the important advantage of creating a "circular economy" system which transforms discarded fibers into useful goods with added value. In addition, ecological benefits related to a lesser use of virgin resources and to the limited need for landfill are achieved [16].

Manufacturing a sustainable building material means controlling the type and amount of energy used for its production, as well as verifying specific requirements on the possibility to recycle or reuse the material at the end of its useful life and on the toxicity of all its components, including the binder together with raw materials [1,2].

In terms of environmental impact of wool recycling, any quantitative assessment is very difficult and affected by high uncertainty, mostly because wool as a virgin material is a co-product with meat. Thus, depending on what we assume to be the main product, the climate impact (estimated by the global warming potential (GWP)) of wool fibers may range from 36.2 kg CO_2 equivalents per kg fibers, to 26 kg CO_2 equivalents per kg fibers [17]. Geographic factors may also significantly affect the above results. By assuming the process of mechanical unravelling of wool to low grade wool yarns to use as a substitute of polyester fibers in blankets, it has been calculated that a significantly positive effect can be obtained [18], compared to incinerating or sending to landfill the wool waste. In particular, in the mentioned study the recycling process assumes that wool may be carded and spun into new yarn to replace other fibers and, consequently, avoiding the largest impact due to the fact that raising of sheep alone accounts for more than 50% of the climate change impacts in the life cycle of wool and about a third of the energy consumption.

With reference to binders, despite the increasing popularity of chitosan in the literature, only one reference [19] could be found discussing its life cycle assessment (cradle to gate). Results showed impressive differences depending on the supply chain, with a climate change impact of 77 kg CO_2 equivalent per kg of chitosan for the European market, and 12.2 kg CO_2 equivalent per kg of chitosan for the Indian supply chain.

Gum Arabic, despite its much older history, and widespread use in the food and drug industry has, to the best of the authors knowledge, never been investigated in terms of environmental impact and life cycle assessment. However, gum Arabic is derived from Acacia plants that grow in fairly dry climates (sub-Saharan Africa and West Asia), and it also contributes to fertilize soil by fixing nitrogen, contrasting desertification. None of the manufacturing processes usually require thermal energy, with transportation and mechanical breaking likely having the largest share of the environmental impact.

In the study reported in this paper, 100% recycled wool composite panels are designed as an alternative to conventional solutions aiming at a sound absorption coefficient of at least 0.5 from 500 Hz on, and a thermal conductivity of at least 0.05 W/(m K). The tested samples were made by using two different natural binders based on chitosan and gum Arabic. Chitosan is a polysaccharide obtained from alkaline deacetylation of chitin from crustacean shells and, thanks to its commercial availability, it is particularly used as a film and adhesive [20]. Several researches [21,22] have been carried out in order to investigate its binding properties. Gum Arabic is a commercially available dried gummy derived from the exudation of steams and branches of Acacia trees. Thanks to its physical and mechanical properties, gum Arabic can be considered a valid substitute for formaldehyde or any other conventional binder in the building materials field [23]. Furthermore, it may be added to cements to improve concrete mechanical properties [24]. The use of organic adhesive solutions to bond recycled post-consumers wool waste allows to produce recyclable materials without further impacts on the environment. The high nitrogen content makes natural fibers biodegradable in a few years unlike synthetic ones [25].

The aim of the research is to characterize the sound absorbing behavior (and other related properties such as thermal insulation) of the innovative tested materials by taking into account, in particular, the effect played by density and porosity variation. The paper also focuses on the study of the suitable placement of the proposed materials by analyzing variation in sound absorption coefficients as a function of the distance from the backing wall. Finally, in order to obtain further elements contributing to material optimization, all the experimental acoustic results were compared with theoretical results based on the empirical prediction model proposed by Delany and Bazley (D&B) [26] and on the phenomenological model suggested by Johnson, Champoux, and Allard (JCA) [27,28].

2. Materials

2.1. Basic Components

Two different binders were used, one from vegetable resources (gum Arabic) and the other one from animal ones (chitosan). Gum Arabic is a hardened sap from the Acacia tree and is commonly used as a natural gum. Its chemical composition is a complex polysaccharide with high molecular weight, water soluble, and its solution gives a slight yellow to reddish color. Gum Arabic is considered a biopolymer. It was purchased from Lana y Telar from Spain.

On the other side, chitosan is also a polysaccharide from crustacean by-products. It is considered a biopolymer and is comprised of partially deacetylated units of 1–4, D-glucosamine. It was purchased from chitosan of medium molecular weight from Sigma-Aldrich. Chitosan is capable of crosslinking itself or by means of a crosslinking agent. In this occasion no crosslinking agent was included. Acetic acid to solubilize chitosan was supplied by Panreac.

Wool is a natural fiber obtained from animals and it is mainly characterized by its proteinic chemical structure and the thermal insulation. It is the natural fiber with the highest limit oxygen index (LOI) which confers resistance against fire ignition. Merino wool fibers used in this study were derived from discarded shreds resulting from the manufacturing process of an Italian clothing company (Gordon Confezioni, Bari, Italy).

2.2. Sample Preparation

The experimented materials were prepared by using 100% Merino wool fibers initially available in the form of cut fabrics (Figure 1a); then, a soft wool batting (Figure 1b) was obtained by carding and by scouring the fibers.

As anticipated, chitosan and gum Arabic were chosen as the natural binders to investigate. Two solutions were achieved by dissolving different amounts of chitosan and gum Arabic in water, as outlined in Table 1. Both blends were mixed in a magnetic stirrer, at room temperature and relative humidity, for about 1 h. The concentration of solutes was chosen, after several attempts, in order to obtain the best combination of porosity and compactness. Finally, materials with potential use as mats with good thermal and acoustic performance were produced.

Two different groups of final composites were produced: The first one was based on textile waste bound with chitosan solution (subsequently referred to as CH); the second type was prepared by binding textile waste with the gum Arabic solution (subsequently referred to as GA).

A three-step process was used to produce the samples. First, the fibers were transformed into wool batting. The second step consisted of soaking the raw materials in the binding solution to let the wool get fully impregnated. The excess amount of liquid was removed by squeezing, mostly to avoid the exceeding binder, that once solid, might compromise a homogeneous distribution of porosity in the final specimen (e.g., by creating surface crusts). In the third step, the achieved mix was compressed into PVC molds to form cylindrical samples. Finally, the samples were dried in an oven at 100 °C, for 1 h and, after drying, they were left in desiccators containing silica gel, under controlled conditions to reach their mass stabilization before testing.

(a) (b)

Figure 1. Textile waste in the form of cut fabrics (**a**) and fleece matrix used as raw material (**b**).

Table 1. Mixing ratio of binding solutions.

Solution	Solute [g]	Water [g]	Acetic Acid [g]
Chitosan	15	1000	5
Gum Arabic	200	1000	-

Several mixes with different bulk densities and same percentages of binder and fibrous matrix were prepared (Table 2). Binder concentration was expressed both as a percentage in wet mass (obtained by weighing the fibers before soaking and after removing the excess binder), and in dry mass (obtained by weighing the final products, after drying them in the oven). Bulk density values are given as mean values among six measurements [29] given that, for each type of binder and for each density value, six cylindrical specimens, 5 cm thick, were prepared (Figure 2). Three were prepared in 10 cm diameter to measure thermal conductivity, thermal diffusivity, specific heat capacity, and normal incidence absorption coefficient at medium-low frequencies. The remaining three samples were prepared in 4 cm diameter to measure normal incidence absorption coefficient at high frequencies and non-acoustic properties (air flow resistance, porosity, and tortuosity).

Table 2. Sample ID, bulk density (with uncertainty given in brackets), and fractional composition for wet and dry samples (in mass).

Sample ID	Bulk Density ρ [kg/m^3]	Fibrous Matrix [wet %]	Binder [wet %]	Fibrous Matrix [dry %]	Binder [dry %]
CH-1	197(1.7)	40	60	95	5
CH-2	145(1.6)	40	60	95	5
CH-3	122(1.2)	40	60	95	5
CH-4	80(1.0)	40	60	95	5
GA-1	177(2.9)	40	60	78	22
GA-2	143(1.6)	40	60	78	22
GA-3	93(1.1)	40	60	78	22

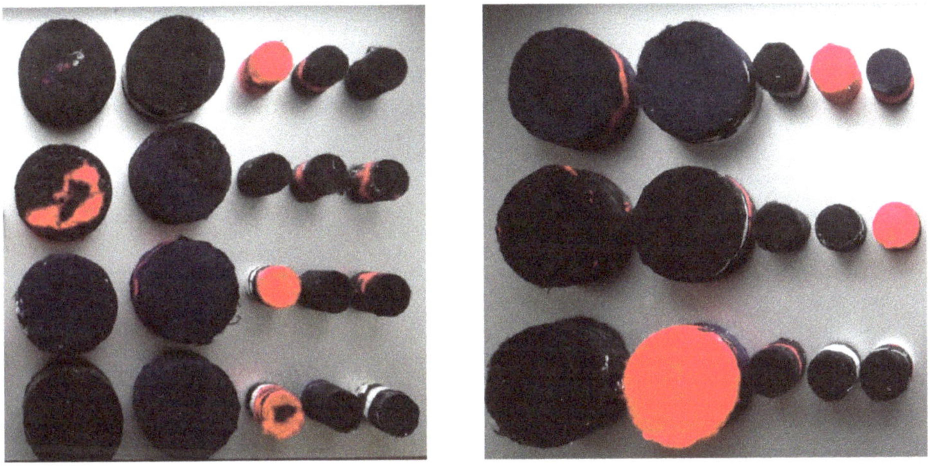

(**a**) (**b**)

Figure 2. Samples with chitosan binder solution (**a**) and samples with gum Arabic binder solution (**b**).

2.3. Microstructural Analysis

A scanning electron microscopy (SEM) analysis was carried out using a Fei Phenom desktop scanner (Thermo Fisher Scientific, Hillsboro, OR, USA). In order to compare the effect of the binder addition on the fibers, samples with and without the binding solution were tested. The samples were prepared by cutting small slices of the fibrous composites onto an adhesive carbon tab attached to an aluminum stub. A sputter coating in gold for a period t = 15 min was applied.

Wool is an animal fiber with keratin as the main protein component (42% in volume). It exhibits a complex and sophisticated structure with the outermost layer (i.e., cuticle) overlapping by scaly cells which are about 10% (in volume) of the fiber. These scales played an important role in the manufacturing process of the building components simplifying the fibers intertwining. During repeated mixing and pressing actions, the scales edges bended, became interlocked and entangled enabling a good cohesion of the resulting fibrous matrix [30]. Figure 3 shows the SEM images of wool fibers without any binder, in which the scale patterns of the cuticle can be observed. According to Patnaik et al. [31], these scales were also effective in dampening the sound wave, especially in the medium high frequency range.

(**a**) (**b**)

Figure 3. Cuticular scale pattern of wool fibers at 500× (**a**) and 2000× (**b**).

Figure 4 shows the SEM images of the fibers bonded with chitosan and gum Arabic solutions. The resulting composite materials were characterized by cylindrical fibers with a 20 μm average diameter, distributed more or less randomly in plane. Both binding solutions joined the fibers preserving the porous matrix of the final composite materials. The fibers intertwined creating a network of tiny air pockets which guaranteed a highly porous product with potential good thermal and acoustic performances.

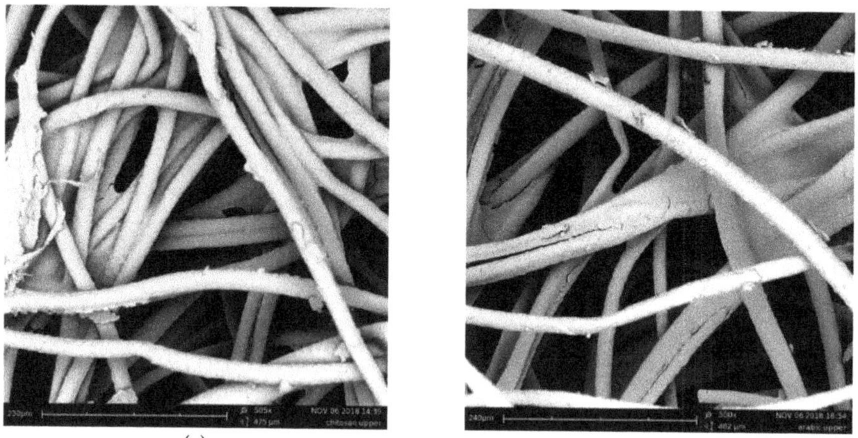

(a) (b)

Figure 4. Wool samples with chitosan binder (**a**) and gum Arabic binder (**b**) at 500×.

3. Methods

3.1. Basic Physical Properties

A ULTRAPYC 1200-e Quantachrome Helium gas pycnometer was used to measure the true density ρ_{true} of the samples. Subsequently, open porosity (ε) was evaluated from the obtained true density and from the bulk density ρ_{bulk}, previously measured according to the gravimetric method. The open porosity of textile fibrous materials corresponds to the inter-fiber and intra-fiber porosity.

$$\varepsilon = 1 - \rho_{\text{bulk}}/\rho_{\text{true}} \tag{1}$$

Tortuosity (τ) was measured in order to evaluate the influence of the pore micro-geometry (i.e., the distribution, the size, and the shape of the pores) on the sound propagation path length in the materials under investigation. As they are non-conducting porous absorbers, the method proposed by Brown [32], based on an electro-acoustic analogy was chosen. According to the authors of [33], three samples for each mix were saturated by soaking in a 10% copper sulfate solution ($CuSO_4$) and, after 24 h, a current was applied using two circular copper plates as electrodes. The electrical resistivity R_0 was calculated for each sample after recording the current intensity values, by varying the voltage between 1 and 8 V. The electrical resistivity of the conducting fluid R_W was also obtained. The tortuosity was given by the following Equation:

$$\tau = \varepsilon\ (R_0/R_W) \tag{2}$$

where ε is the porosity of the sample.

The tortuosity was also empirically estimated according to Berryman's formula [34]:

$$\tau_{est} = 1 + (1 - \varepsilon)/2\varepsilon \tag{3}$$

Air flow resistance (σ_s) of three samples from each mix was measured according to the method proposed by Ingard and Dear [35] which uses sound waves. Although this procedure could not be considered as steady as the standardized approach, del Rey et al. [36] proved that it was scientifically accurate when adopted for thin or scarcely resistive samples, as occurs in the present research.

The measurement setup consisted of a 5 mm thick methacrylate tube, with a 4 cm inner diameter. The tube was divided in two parts; each of them 85 cm long. At one end there was a 5 cm loudspeaker (Visaton FRS 5, Visaton, Haan, Germany) with a frequency response spanning from 150 to 20 kHz; at the other end there was a rigid termination made by a 5 cm thick methacrylate. The tested samples were assembled between the two tubes. The flow resistance was deduced using the transfer function of the two microphones excited by an exponential sine sweep. All of the processing was performed by a custom-made MATLAB® (2018, Mathworks, Natick, MA, USA) graphic user interface.

Thermal performances were measured, on three samples for each typology, using a transient plane source apparatus, ISOMET 2104 (Applied Precision, Bratislava, Slovakia). This device was used by several authors [37,38] in order to test the thermal behavior of fibrous materials. The measurement was based on the analysis of temperature response of each sample to heat flow impulses [39]. Prior to testing the thermal properties, all the specimens were oven dried at 105 °C then cooled to ambient temperature in desiccators containing silica gel. The dry-state thermal conductivity λ_{dry} was then measured.

3.2. Acoustic Properties

According to ISO 10534-2:1998 [40], sound absorption measurements were performed by the transfer function method. Two tubes with different diameters (10 and 4 cm) and a thickness of 5 mm were used for the test with the aim to consider the largest spectrum range. The tube with an internal diameter of 10 cm had a maximum measurable frequency of 2 kHz and it used two different microphone distances (6 and 20 cm, respectively yielding a low frequency limit of 400 and 50 Hz). The emitting end consisted of an 11 cm loudspeaker sealed into a wooden case and suitably isolated from the tube structure by an elastic and protective layer. The second tube, with a diameter of 4 cm, was the same used for the flow resistance measurement. In this case the microphone spacing was 3 cm, and the frequency covered a range between 200 and 5 kHz. All the processing was performed by a MATLAB® (2018, Mathworks, Natick, MA, USA) graphic user interface generating a 5 s linear sweep from 70 to 3 kHz, used in combination with the largest tube, and from 500 to 5 kHz considering the smallest tube.

The measured sound absorption coefficients were compared with those estimated by theoretical prediction models. The empirical model proposed by Delany and Bazley [26] and the phenomenological model developed by Johnson, Champoux, and Allard [27,28] were chosen. In both situations, the absorption coefficients of the investigated fibrous materials were obtained after determining the characteristic impedance Zc and the wavenumber k. The empirical model just links the sound absorption coefficients to the air flow resistivity σ (i.e., the air flow resistance per unit thickness) of the materials. Thus, to solve the D&B equations only one non-acoustic property was used to calculate the parameter $X = (\rho_0 f)/\sigma$ (f being the frequency and ρ_0 being the density of the air), needed to determine the characteristic impedance Z_C and the wavenumber k:

$$Z_C = \rho_0 c_0\,(1 + 0.057X^{-0.754} - j0.087X^{-0.732}) \tag{4}$$

$$k = (\omega/c_0)\,(1 + 0.0978\,X^{-0.700} - j0.189X^{-0.595}) \tag{5}$$

where c_0 is the speed of sound in air and ω is angular frequency. The D&B equations can be considered valid only in a defined frequency range given by $0.01 < X < 1.0$, and for σ values below 50 kN·s/m^4 [41].

Although the D&B empirical model has been successfully tested over a variety of fibrous materials having porosity close to unity (similar to those under investigation), it nonetheless neglects the important effects that other structural parameters of the materials may have to influence their acoustic performance. For this reason, the JCA model was also considered, according to which Z_C and k are expressed as a function of the effective (or dynamic) bulk density ρ_e and bulk modulus k_e:

$$Z_C = (k_e\ \rho_e)^{0.5} \tag{6}$$

$$k = (\rho_e/k_e)^{0.5} \tag{7}$$

The computation of these two parameters involves the use of other physical quantities including porosity ε, tortuosity τ, viscous characteristic length Λ, and thermal characteristic length Λ', in addition to flow resistivity and density. Most of such coefficients (porosity, tortuosity, flow resistivity, and density) were analyzed in Section 3.1. A more extensive explanation of the physical meaning of the two characteristic lengths is beyond the scope of the paper, so reference to the original paper [28] or textbooks [41] is suggested. Briefly, the two characteristic lengths describe the effects of viscosity and thermal dissipative forces inside the porous structure. The viscous characteristic length is the weighted ratio of the volume to surface area of the pores. It takes into account some significant parameters, i.e., the pore shape represented by a variable s which reflects the deviation of pore shape from an ideal circle and lies between 0.3 and 3. The thermal characteristic length is needed for materials showing a complex internal structure. Usually, $\Lambda' \geq \Lambda$ and, in first approximation, $\Lambda' = 2\ \Lambda$.

3.3. *Ignitability Test*

An ignitability test was carried out, as far as possible according to ISO 11925-2:2010 [42]. The major non-standard compliant element was represented by the shape and dimension of the sample that, for this batch, was only produced in cylindrical molds having 10 cm diameter and 5 cm thickness. Thus, an "edge exposed" test was carried out, with a 15 s flame application, on two different samples of the selected typology. A reference sample, made of pressed wool fibers (3 cm thick) was used as a reference for comparison purposes, together with GA-3 and CH-4 samples.

4. Results

4.1. *Basic Physical Properties*

The results of the non-acoustic parameters are given in Table 3. The values of porosity, tortuosity, air flow resistance, and air flow resistivity are given as mean values of the experimental data together with their measurement uncertainty calculated according to Reference [29]. Air flow resistivity (i.e., the air flow resistance per unit thickness) was also calculated as it is more frequently used as a criterion for choosing suitable materials for noise control applications. As seen in Section 3.2, it also plays an important role as an input parameter in prediction models.

Table 3. Summary of non-acoustic parameters for each sample.

Sample Code	Bulk Density, ρ [kg/m^3]	Porosity, N [-]	Measured Tortuosity, τ [-]	Estimated Tortuosity, τ_{est} [-]	Air Flow Resistivity, σ [kN·s/m^4]	Thermal Conductivity, λ [W/(m·K)]
CH-1	197(1.7)	0.86(0.002)	1.23(0.07)	1.08(0.0028)	66.3(7.8)	0.060(0.0005)
CH-2	145(1.6)	0.89(0.001)	1.11(0.05)	1.06(0.0008)	46.0(9.1)	0.055(0.0007)
CH-3	122(1.2)	0.91(0.001)	1.12(0.05)	1.05(0.0013)	16.5(2.7)	0.052(0.0006)
CH-4	80(1.0)	0.94(0.001)	1.20(0.18)	1.03(0.0016)	11.7(2.2)	0.049(0.0005)
GA-1	177(2.9)	0.87(0.0014)	-	1.08(0.0019)	44.7(5.0)	0.059(0.0012)
GA-2	143(1.6)	0.90(0.0003)	-	1.06(0.0004)	23.6(2.5)	0.056(0.0006)
GA-3	93(1.1)	0.93(0.0004)	-	1.04(0.0006)	14.4(2.3)	0.050(0.0005)

As shown in Table 3, all materials were characterized by a porosity close to 0.9, regardless of the nature of the binder. This was in agreement with the porosity values of felts and mineral fiber materials ranging respectively from 0.83 to 0.95 and from 0.92 to 0.99 [43]. A void fraction close to 90% was expected in view of the microstructural analysis carried out in Section 2.3.

Tortuosity could not be measured for GA samples because gum Arabic is soluble in water and this would have prevented a suitable placement of the samples in the measurement equipment. Thus, in this case τ was only estimated according to Equation (3). A comparison between measured and predicted values, both available for CH samples, showed that predicted values were slightly underestimated with differences between 5% and a maximum of 15%. Thus, similar differences might be expected in the GA samples. In both cases, tortuosity values are in agreement with close-to-unity values typical of fibrous materials, where the fluid always takes the straightest path due to the absence of solid grains [44]. Considering that tortuosity may assume much larger values (e.g., up to 4 for densely packed granular materials), a 10% discrepancy may be considered negligible and, more importantly, of no significant effect on the sound absorption coefficients (as shown in next section).

As confirmed by results reported in Table 3, air flow resistivity for all tested materials was below 100 kN·s/m^4, which is the limiting value for a material to be considered as an impervious layer, as reported by del Rey et al. [45], and only in one case it exceeded 50 kN·s/m^4 corresponding to the limiting value for the application of D&B model. For all samples air flow resistivity proved to be well correlated to bulk density values (Figure 5), as demonstrated by the statistically significant ($R^2 = 0.8891$) exponential correlation that was found.

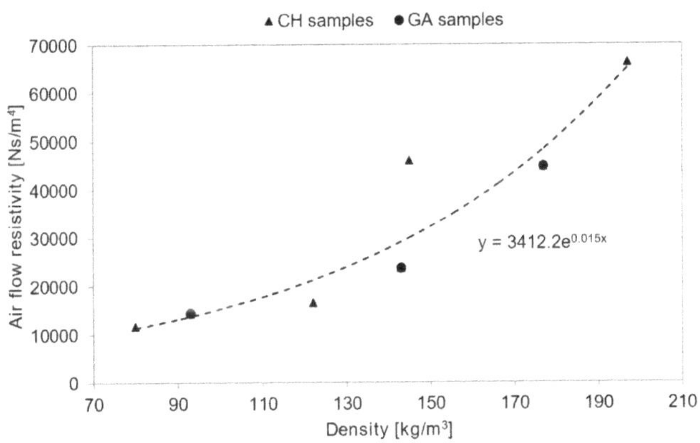

Figure 5. Experimental air flow resistivity versus density.

The experimental results obtained for non-acoustic properties of the studied materials were in agreement with some vegetal woolen building materials with similar density values [46,47]. Glé et al. [46] carried out a research about the acoustic behavior of vegetal woolen materials and showed that these biobased panels had high porosity and a tortuosity of about one, similar to more conventional fibrous composites. Additionally, an increase of the air flow resistivity values with the increase in density was observed. A similar dependence on density was found for the effective thermal conductivity. It can be noted that both CH and GA samples showed similar thermal conductivity values as a function of the bulk density (Table 3). CH samples had λ varying between 0.049 and 0.060 W/(m K) with a bulk density ρ ranging from 80 to 197 kg/m^3, while for GA samples, λ varied from 0.050 to 0.059 W/(m K) with bulk density ρ ranging between 93 and 177 kg/m^3.

Figure 6 shows the thermal conductivity values of CH and GA samples as a function of their density values. Statistically significant linear correlations were found for both samples ($R^2 = 0.997$ for CH and $R^2 = 0.999$ for GA). Similar linear trends were observed for the two groups of materials, with λ decreasing linearly when the density dropped. However, a statistically significant difference between the two regressions was found (with a residual probability $p < 0.005$), resulting in CH samples having a thermal conductivity about 7% lower than a same-density GA sample. This small difference does not seem to be related to the sample porosity, which was very similar for same-density samples, neither could be better explained by the analysis of the microstructure. In fact, the SEM images (Figure 3) demonstrated that the use of different binders barely affected the porous matrix of the final composite (Table 3). Conversely, the different fraction of binder (on the dry sample) outlined in Table 2, may well explain the 7% variation in thermal conductivity. In fact, GA samples in which the same amount of wool fibers is bound by a 22% in mass of binder, against the 5% of CH samples.

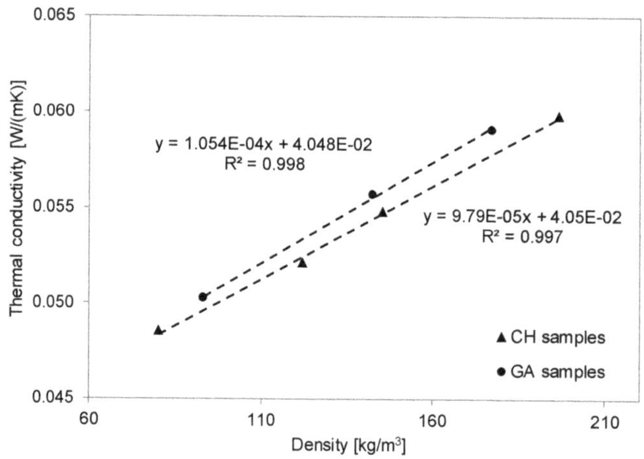

Figure 6. Thermal conductivity λ_{dry} versus bulk density ρ for chitosan (CH) and gum Arabic (GA) samples.

4.2. Acoustic Properties

The characterization of the acoustic performance was carried out by comparing the normal incidence absorption coefficients measured in the laboratory with those predicted using Delany–Bazley and Johnson–Champoux–Allard equations. Data used to feed the D&B model, i.e., air flow resistivity, were derived from experimental measurements (Table 3). The calculation of JCA predictions was carried out using some previously measured parameters, i.e., density, porosity, tortuosity, and air flow resistivity (Table 3), while the values for the shape factor and the characteristic length ratio Λ'/Λ were numerically adjusted to get the best agreement between measured and predicted values. The resulting set of parameters is given in Table 4. A comparison between results obtained using measured and estimated tortuosity confirmed that differences in the relevant mean absolute errors were usually negligible, thus only estimated values were used in all the cases.

Table 4. Coefficients used to feed the Johnson–Champoux–Allard (JCA) model.

Sample Code	Shape Cactor, s [-]	Ratio, Λ'/Λ [-]	Mean Absolute Error [-]
CH-1	2.0	2	0.0172
CH-2	0.8	2	0.0240
CH-3	3.0	2	0.0295
CH-4	2.0	2	0.0334
GA-1	1.5	2	0.0307
GA-2	2.5	2	0.0304
GA-3	2.0	2	0.0263

Figure 7 shows the normal incidence sound absorption for CH-4 and GA-3 samples, both having low and comparable air flow resistance values. As it can be observed, the experimental curves of the two materials were almost overlapping and both trends were in accordance with those estimated by the theoretical approaches. In particular, the JCA model allowed a rather precise estimation of peak placement, its maximum–minimum fluctuations and values of absorption coefficients. The phenomenological model perfectly predicted the first peak appearing around 1250 Hz with α close to unity and the following drop around 2500 Hz. It is important to point out that optimization of the estimated parameters had a negligible effect on the final result.

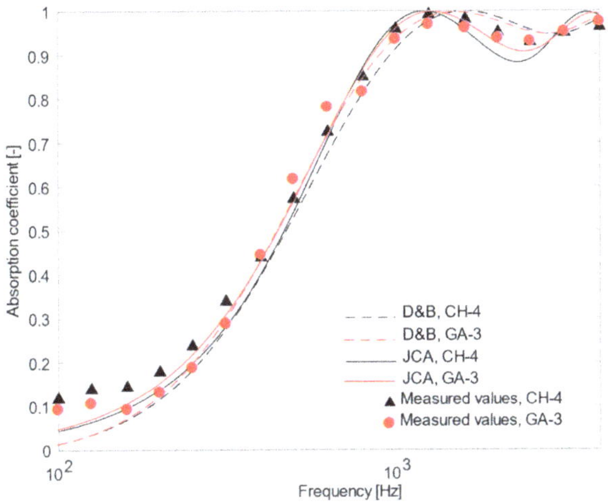

Figure 7. Plot of mean measured normal incidence absorption coefficients and those predicted using Delany and Bazley and Johnson–Champoux–Allard equation for CH-4 sample (in black) and GA-3 sample (in red).

Figure 8 shows a comparison between CH-3 and GA-2 samples. The curves of the two samples were similar up to 630 Hz, where some differences could be noted. For GA-2 a first mild peak appeared around 630 Hz, followed by another peak around 1250 Hz, almost coinciding with the first peak of CH-3 sample. The measured values showed good agreement from 1600 Hz on. Considering the CH-3 case, the phenomenological model performed better than the empirical one. The first peak predicted by JCA model, although overestimated, was almost aligned to the measured one. Taking into account the GA-2 material, significant discrepancies appeared between 600 and 1600 Hz where measured values were lower than expected according to the models, and apparently no combination of parameters could return a similar behavior. Thus, the discrepancy should suggest that some anomalous behavior took place with GA-2 samples.

A comparison between Figures 7 and 8 pointed out that the use of different binders played a negligible role in changing the acoustic performance of the compared samples (CH-3, CH-4, GA-2, and GA-3), while the structure of the materials mainly affected their acoustic behavior. All the samples, apart from limited exceptions, showed sound absorption coefficients which were consistent with that of porous materials of the same thickness and having the corresponding flow resistance values (ranging between 585 and 1182 Ns/m^3). However, the fact that JCA model showed a better agreement than D&B confirmed the importance of the microstructure in altering the sound path and, for example, moving the peak of absorption towards lower frequencies (compared to the expected theoretical peak which should appear at the frequency having the quarter-wavelength equal to the sample thickness). As shown in Table 4, the shape factor estimated for CH-3, CH-4, GA-2, and GA-3 samples was larger than one (ranged between 2 and 3), demonstrating that despite tortuosity being close to unity, the fibers are arranged according to an intricate internal structure, which is far from the "ideal" cylindrical pores.

Figure 9 shows the sound absorption curves for the samples with higher flow resistance (CH-2 and GA-1). Clearly, the absorption coefficients differed significantly from those previously analyzed, as they increased at low frequencies and decreased above 800 Hz. At 315 Hz, α was about 0.5 for GA-1 and CH-2 samples against a value of about 0.3 for CH-4 and GA-3 samples and 0.4 for CH-3 and GA-2 samples. At 1250 Hz, α was about 0.8 for GA-1 and CH-2 samples against a value of about 1 for CH-4 and GA-3 samples and 0.9 for CH-3 and GA-2 samples. This was because samples CH-2 and GA-1 had high air flow resistivity values (46 and 44.7 kN·s/m^4, respectively) that increased the viscous and thermal interaction inside the micro pores forming the material, while increasing, at the same time, the surface impedance.

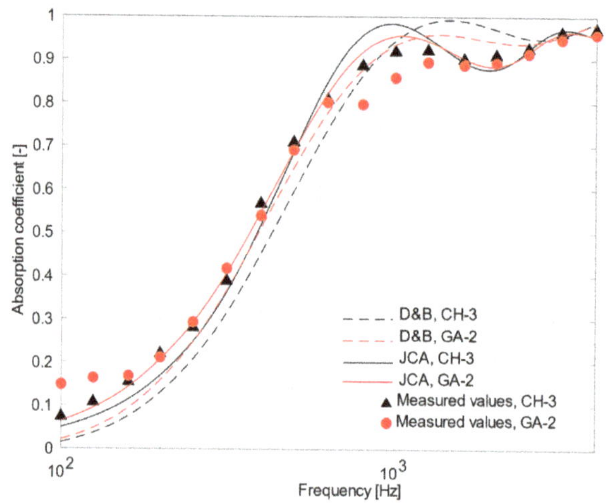

Figure 8. Plot of mean measured normal incidence absorption coefficients and those predicted using Delany and Bazley and Johnson–Champoux–Allard equation for CH-3 sample (in black) and GA-2 sample (in red).

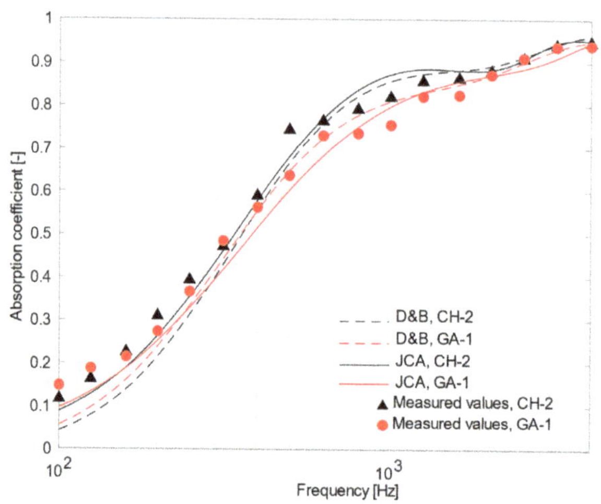

Figure 9. Plot of mean measured normal incidence absorption coefficients and those predicted using Delany and Bazley and Johnson–Champoux–Allard equation for CH-2 sample (in black) and GA-1 sample (in red).

As a further confirmation of this behavior, Figure 10 shows the absorption curves referred to the sample CH-1, having the highest density and the highest air flow resistivity values. The absorption coefficients decreased significantly at high frequencies, reaching $\alpha = 0.73$ at 1250 Hz. Such expected behavior was due to a higher flow resistivity value, higher than 60 kN·s/m^4. It is worth noting that D&B performed fairly well, despite the flow resistivity was, in this case, above the maximum recommended value.

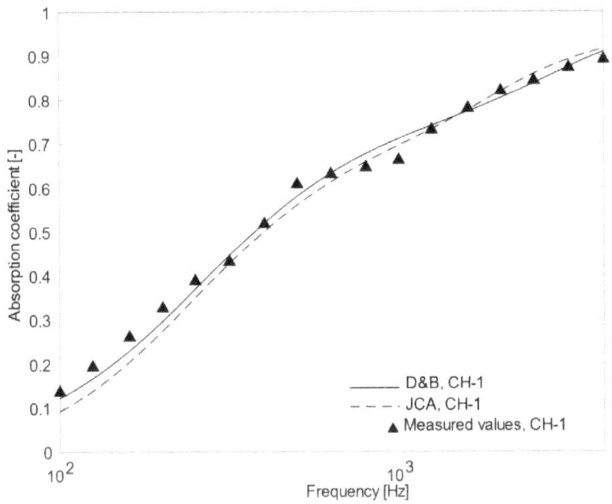

Figure 10. Plot of mean measured normal incidence absorption coefficients and those predicted using Delany and Bazley and Johnson–Champoux–Allard equation for CH-1 sample.

As confirmed by Figures 9 and 10, both phenomenological and empirical models estimated the acoustic behavior of the denser materials characterized by a higher air flow resistivity values (CH-1, CH-2, and GA-1) almost with the same good accuracy. On the contrary, the JCA model outperformed the D&B model in comparison with measured absorption coefficients of the less dense materials (CH-3, CH-4 and GA-2, GA-3) with lower air flow resistivity values. Likely, when flow resistivity increases, the two models tend to converge, because the tighter packing of the fibers makes resistivity play a major role also in the JCA model.

4.3. *Effect of Air Gap*

The effect of the air gap between the porous samples and the rigid backing was investigated by analyzing different placements of the sound absorbing materials under test. In view of what has been concluded in Section 4.3, theoretical models allowed a more accurate prediction of the experimental acoustic behavior of materials with a higher porosity, for which JCA model was more accurate than D&B. Thus, the more porous samples, CH-4 and GA-3, were tested by placing them in the impedance tube with 50, 100, and 150 mm air gaps and only the phenomenological model was used as comparison.

It is well known that placing a porous material at a distance from a rigid surface is a valid technique to improve sound absorption at low frequencies, as an alternative to increasing the thickness of the material. In fact, Figures 11 and 12 show that the peak value of the sound absorption curves measured for the two samples shifted towards lower frequencies as the air gap increased. The minimum, appearing at twice the frequency where the maximum is located, tends to be deeper when the air gap increases, but under diffuse field conditions a much smoother response is expected. Finally, it is worth noticing that, even under these conditions, the agreement between measured values and those predicted using the JCA model remains very good.

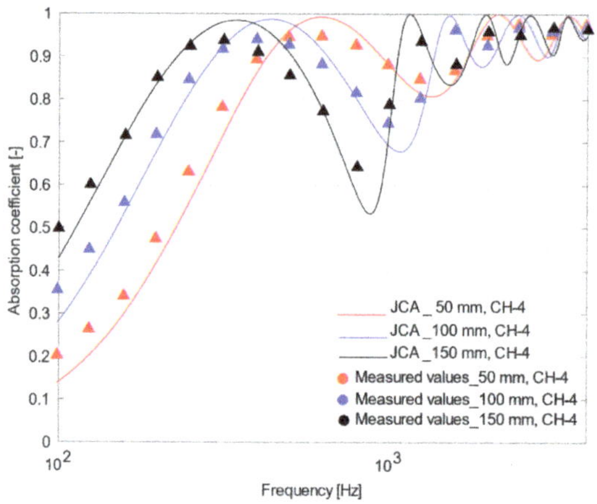

Figure 11. Plot of mean measured normal incidence absorption coefficients and those predicted using Johnson–Champoux–Allard equation for CH-4 sample placed at 50, 100, and 150 mm from the rigid surface.

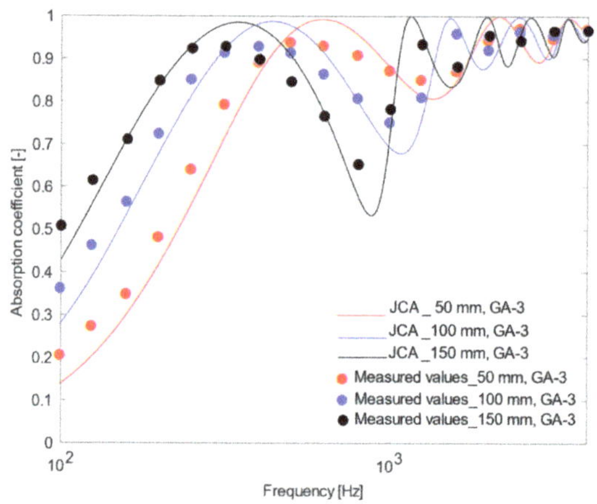

Figure 12. Plot of mean measured normal incidence absorption coefficients and those predicted using Johnson–Champoux–Allard equation for GA-3 sample placed at 50, 100, and 150 mm from the rigid surface.

4.4. Ignitability Test

Results of the ignitability test are given in Figure 13 and show some interesting differences. It can be observed that pressed wool (Figure 13a,d) presented the best behavior, with a very limited flame propagation on the flat face (not exceeding 60 mm from ignition point), some smoke, and no droplets of melted material. Flame extinguished immediately after the burner was retracted. Samples with chitosan binder (Figure 13b,e) showed a quite different behavior, with a flame spreading up to the topmost part of the sample, limited smoke, and no droplets. The flame extinguished as soon as the burner was retracted, but very small carbonized portions kept burning for about 5 s, with smoke production. Finally, samples with gum Arabic binder showed a limited flame spreading on the flat surface (not exceeding 80 mm from the ignition point), copious smoking, and no droplets. No flaming and smoking was observed after the burner was retracted. Thus, the test pointed out significant differences that are worth being further investigated, but results are promising, also in the light of potentially improving this behavior by means of additives mixed to the binder [48].

Figure 13. Samples used during ignitability test before (**a**–**c**) and after (**d**–**f**) flame application, (**a**,**d**) correspond to pressed wool fibers (no binders), (**b**,**e**) correspond to recycled wool with chitosan binder (CH-4), and (**c**,**f**) correspond to recycled wool fibers with gum Arabic binder (GA-3).

5. Practical Implications and Limitations

The scope of the present paper was that of investigating the performance of composite materials obtained from the use of fibers derived from recycled textile wastes (100% merino wool), combined with different binders to be used as building materials with sound absorbing and thermal insulating properties. Results that were presented in the previous section confirmed that such materials have similar or better performance than conventional thermal insulating and sound absorbing materials. Thus, in order to understand whether they might be actually competitive, worth being used in the real world, and hence being industrialized in some way, it is important to answer to some questions:

Do they provide comparable resistance to aging than conventional products? Do they provide comparable fireproof ratings as conventional materials? Additionally, above all, is their environmental impact lower, or equal, than conventional materials?

Clearly, providing a specific answer to all of these questions would require further studies which are under way. However, it is possible to make some preliminary considerations based on literature evidences. With reference to aging and resistance to external attacks, wool is known to have a very good capacity for storing water vapor, thus preventing the material from getting damp. Consequently, wool naturally offers good resistance to mold and fungi. In addition, chitosan has been proved to contribute to further prevent mold and bacteria formation [49], and similar antibacterial effects have been shown by gum Arabic [50].

In terms of fire resistance wool is known to outperform any other textile fiber (both natural and synthetic), because it has a very high ignition temperature of 570–600 °C combined with a high limiting oxygen index (that measures the amount of oxygen needed to sustain combustion) and low combustion heat. Thus, in the light of the above features it is self-extinguishing. In addition, wool does not melt and fibers swell when heated, creating a tighter layer that prevents flame from spreading. However, as demonstrated by ignition tests, binders affected negatively such performance, particularly in terms of flame spreading (for chitosan) and smoke generation (for gum Arabic). Therefore, it becomes of primary importance to investigate either the chemical processes behind combustion and the use of alternative binders or flame retardant treatments.

In terms of environmental impact of the binders, the figures that were presented in the introduction might be properly "rescaled" to take into account the actual amounts of materials that are required, based on the sample composition. Therefore, as the binding solution includes 15 g chitosan per kg water, and that the amount of binder is 60% the mass of the sample, this results in just 9 g of chitosan per kg of (wet) sample prepared. Thus, the higher environmental costs of chitosan production are only marginally reflected on the panel preparation. With reference to gum Arabic the situation is a bit different as the amount of dry product requested to manufacture 1 kg of composite panel is 120 g, which is much higher than chitosan. However, in the light of the positive effects related to Acacia tree cultivation, we might expect a lower overall impact.

6. Conclusions

The experimental analysis of the thermal properties showed that all the samples have thermal conductivity ranging between 0.049 and 0.060 W/(m K), independent of the binder used. As expected, λ values linearly increased with increasing materials density and decreasing material porosity.

Porosity variation also affected the air flow resistivity of samples, influencing their acoustic performance. It was observed that more porous samples were characterized by a lower air flow resistivity, showing a better sound absorption in the mid and high frequency ranges (with α higher than 0.8 at frequencies above 500 Hz). On the contrary, the increased air flow resistivity values of the less porous samples improved sound absorption at low frequencies yielding α as high as 0.5 from 315 Hz on.

A comparison between measured and predicted absorption coefficients was useful to identify the Johnson, Champoux, and Allard model as the most suitable to perform any optimization exercise. In addition, using some of the selected samples in combination with air gaps allowed achieving α higher than 0.8 at frequencies above 315 Hz, using only a 5 cm gap.

Further investigations are under way in order to define the mechanical characteristics of the samples and the practical application of the proposed solutions in the building industry. In addition, given the substantial equivalence between chitosan and gum Arabic in terms of performance, a life cycle assessment will be carried out to better clarify which solution is more eco-friendly.

Author Contributions: Conceptualization, C.R., M.B.A., S.L., P.S., and F.M.; Methodology, C.R., M.B.A., S.L., P.S., and F.M.; Investigation, C.R., M.B.A., S.L., P.S., and M.Z.C.; Formal analysis, C.R., S.L., P.S., F.M.; Data curation, C.R.; Writing—original draft preparation, C.R.; Writing—review and editing, C.R., M.B.A., J.G.-P., S.L., P.S., and F.M.; Visualization, C.R.; Software, F.M.; Supervision, P.S. and F.M.; Funding acquisition, P.S. and F.M.

Funding: C.R. scholarship has been funded by the Italian Ministry of Education, University and Research (MIUR), within the National Research Program "PON Ricerca e Innovazione 2014-2020" (grant DOT1748713 N.5).

Acknowledgments: Authors acknowledge the staff at Electron Microscopy Service of the Universitat Politècnica de València for their support on the analysis of the samples. Authors wish to address a special thanks to the Company Gordon Confezioni srl (Cassano, Italy) for supplying the raw materials and for the contribution given to this research.

Conflicts of Interest: The authors declare no conflict of interest.

References

1. Joshi, S.V.; Drzal, L.T.; Mohanty, A.K.; Arora, S. Are natural fiber composites environmentally superior to glass fiber reinforced composites? *Compos. Part A Appl. Sci. Manuf.* **2004**, *35*, 371–376. [CrossRef]

2. Hesterberg, T.W.; Hart, G.A. Synthetic Vitreous Fibers: A Review of Toxicology Research and Its Impact on Hazard Classification. *Crit. Rev. Toxicol.* **2001**, *31*, 1–53. [CrossRef]
3. Bakatovich, A.; Davydenko, N.; Gaspar, F. Thermal insulating plates produced on the basis of vegetable agricultural waste. *Energy Build.* **2018**, *180*, 72–82. [CrossRef]
4. Martellotta, F.; Cannavale, A.; De Matteis, V.; Ayr, U. Sustainable sound absorbers obtained from olive pruning wastes and chitosan binder. *Appl. Acoust.* **2018**, *141*, 71–78. [CrossRef]
5. Kymalainen, H.R.; Sjoberg, A.M. Flax and hemp fibres as raw materials for thermal insulations. *Build. Environ.* **2008**, *43*, 1261–1269. [CrossRef]
6. Zhou, X.Y.; Zheng, F.; Li, H.G.; Lv, C.L. An environment-friendly thermal insulation material from cotton stalk fibers. *Energy Build.* **2010**, *42*, 1070–1074. [CrossRef]
7. Ashour, T.; Georg, H.; Wu, W. Performance of straw bale wall: A case of study. *Energy Build.* **2011**, *43*, 2518–2523. [CrossRef]
8. Lim, Z.Y.; Putra, A.; Nor, M.J.M.; Yaakob, M.Y. Sound absorption performance of natural kenaf fibres. *Appl. Acoust.* **2018**, *130*, 107–114. [CrossRef]
9. Fouladi, M.H.; Ayub, M.; Nor, M.J.M. Analysis of coir fiber acoustical characteristics. *Appl. Acoust.* **2011**, *72*, 35–42. [CrossRef]
10. Wei, K.; Lv, C.; Chen, M.; Zhou, X.; Dai, Z.; Shen, D. Development and performance evaluation of a new thermal insulation material from rice straw using high frequency hot-pressing. *Energy Build.* **2015**, *87*, 116–122. [CrossRef]
11. Berardi, U.; Iannace, G. Acoustic characterization of natural fibers for sound absorption applications. *Build. Environ.* **2015**, *94*, 840–852. [CrossRef]
12. Rubino, C.; Liuzzi, S.; Stefanizzi, P.; Martellotta, F. Textile wastes in building sector: A review. *Model. Meas. Control B* **2018**, *87*, 172–179. [CrossRef]
13. Ricciardi, P.; Belloni, E.; Cotana, F. Innovative panels with recycled materials: Thermal and acoustic performance and life cycle assessment. *Appl. Energy.* **2014**, *134*, 150–162. [CrossRef]
14. Del Mar Barbero-Barrera, M.; Pombo, O.; de los Angeles Navacerrada, M. Textile fibre waste bindered with natural hydraulic lime. *Compos. Part B* **2016**, *94*, 26–33. [CrossRef]
15. Echeverria, C.A.; Pahlevani, F.; Handoko, W.; Jiang, C.; Doolan, C.; Sahajwalla, V. Engineered hybrid fibre reinforced composites for sound absorption building applications. *Resour. Conserv. Recy.* **2019**, *143*, 1–14. [CrossRef]
16. Leal Filho, W.; Ellams, D.; Han, S.; Tyler, D.; Boiten, V.J.; Paço, A.; Moora, H.; Balogun, A. A review of the socio economic advantages of textile recycling. *J. Clean. Prod.* **2019**, *218*, 10–20. [CrossRef]
17. Sandin, G.; Roos, S.; Johansson, M. *Environmental Impact of Textile Fibers-What We Know and What We Don't Know*; MISTRA Future Fashion report 2019:03 part 2; Rise AB: Göteborg, Sweden, 2019.
18. Schmidt, A.; Watson, D.; Roos, S.; Askham, C.; Brunn Poulsen, P. *Gaining benefits from discarded textiles*; Rosendahls-Schultz Grafisk: Hillerød, Denmark, 2016; pp. 107–110.
19. Munoz, I.; Rodriguez, C.; Gillet, D.; Moerschbacher, B.M. Life cycle assessment of chitosan production in India and Europe. *Int. J. Life Cycle. Assess.* **2018**, *23*, 1151–1160. [CrossRef]
20. Mati-Baouche, N.; De Baynast, H.; Lebert, A.; Sun, S.; Lopez-Mingo, C.J.S.; Leclaire, P.; Michaud, P. Mechanical, thermal and acoustical characterizations of an insulating bio-based composite made from sunflower stalks particles and chitosan. *Ind. Crops. Prod.* **2014**, *58*, 244–250. [CrossRef]
21. El Hage, R.; Khalaf, Y.; Lacoste, C.; Nakhl, M.; Lacroix, P.; Bergeret, A. A flame retarded chitosan binder for insulating miscanthus/recycled textile fibers reinforced biocomposites. *J. Appl. Polym. Sci.* **2019**, *136*, 47306. [CrossRef]
22. Mati-Baouche, N.; de Baynast, H.; Michaud, P.; Dupont, T.; Leclaire, P. Sound absorption properties of a sunflower composite made from crushed stem particles and from chitosan bio-binder. *Appl. Acoust.* **2016**, *111*, 179–187. [CrossRef]
23. Abuarra, A.; Hashim, R.; Bauk, S.; Kandaiya, S.; Taghizadeh Tousi, E. Fabrication and characterization of gum Arabic bonded Rhizophora spp. Particleboards. *Mater. Des.* **2014**, *60*, 108–115. [CrossRef]
24. Elinwa, A.U.; Abdulbasir, G.; Abdulkadir, G. Gum Arabic as an admixture for cement concrete production. *Constr. Build. Mater.* **2018**, *176*, 201–212. [CrossRef]
25. Baker, I. Wool. In *Fifty Materials That Make the World*; Springer: New York, NY, USA, 2018; pp. 261–265.

26. Delany, M.E.; Bazley, E.N. Acoustical properties of fibrous materials. *Appl. Acoust.* **1970**, *3*, 105–116. [CrossRef]
27. Johnson, D.L.; Koplik, J.; Dashen, R. Theory of dynamic permeability and tortuosity in fluid-saturated porous media. *J. Fluid Mech.* **1987**, *176*, 379–402. [CrossRef]
28. Allard, J.F.; Champoux, Y. New empirical equation for sound propagation in rigid frame fibrous material. *J. Acoust. Soc. Am.* **1992**, *91*, 3346–3353. [CrossRef]
29. *ISO/IEC Guide 98-3:2008 (JCGM/WG1/100) Uncertainty of Measurement—Part 3: Guide to the Expression of Uncertainty in Measurement (GUM:1995)*; ISO: Geneva, Switzerland, 2008.
30. Tridico, S.R. Natural animal textile fibres: Structure, characteristics and identification. In *Identification of Textile Fibers*; Houck, M.M., Ed.; Woodhead Publishing in Textiles: New York, NY, USA, 2009; pp. 27–67.
31. Patnaik, A.; Mvubu, M.; Muniyasamy, S.; Botha, A. Thermal and sound insulation materials from waste wool and recycled polyester fibers and their biodegradation studies. *Energy Build.* **2015**, *92*, 161–169. [CrossRef]
32. Brown, J.S. Connection between formation factor for electrical resistivity and fluid-solid coupling factors in Biot's equations for acoustic waves in fluid-filled porous media. *Geophysics* **1980**, *45*, 1269–1275. [CrossRef]
33. Pfretzschner, J.; Rodriguez, R.M. Acoustic properties of rubber crumbs. *Polym. Test.* **1999**, *18*, 81–92. [CrossRef]
34. Berryman, J.G. Confirmation of Biot's theory. *Appl. Phys. Lett.* **1980**, *37*, 382–384. [CrossRef]
35. Ingard, U.K.; Dear, T.A. Measurement of acoustic flow resistance. *J. Sound Vib.* **1985**, *103*, 567–572. [CrossRef]
36. Del Rey, R.; Alba, J.; Arenas, J.P.; Ramis, J. Evaluation of Two Alternative Procedures for Measuring Air Flow Resistance of Sound Absorbing Materials. *Arch. Acoust.* **2013**, *38*, 547–554.
37. Jerman, M.; Černý, R. Effect of moisture content on heat and moisture transport and storage properties of thermal insulation materials. *Energy Build.* **2012**, *53*, 39–46. [CrossRef]
38. Jerman, M.; Palomar, I.; Kočí, V.; Černý, R. Thermal and hygric properties of biomaterials suitable for interior thermal insulation systems in historical and traditional buildings. *Build. Environ.* **2019**, *154*, 81–88. [CrossRef]
39. Gustafsson, S.E. Transient plane source techniques for thermal conductivity and thermal diffusivity measurements of solid materials. *Rev. Sci. Instrum.* **1991**, *62*, 797. [CrossRef]
40. *ISO 10534-2, Acoustics—Determination of Sound Absorption Coefficient and Impedance in Impedance Tubes—Part 2: Transfer-Function Method*; ISO: Geneva, Switzerland, 1998.
41. Cox, T.J.; D'Antonio, P.K. *Acoustic Absorbers and Diffusers, Theory, Design and Application*; Spon Press: London, UK, 2004; pp. 141–145.
42. *ISO 11925-2:2010, Reaction to Fire Tests—Ignitability of Products Subjected to Direct Impingement of Flame—Part 2: Single-Flame Source Test*; ISO: Geneva, Switzerland, 2010.
43. Mechel, F.P. Porous Absorbers. In *Formulas of Acoustics*; Mechel, F.P., Ed.; Springer: New York, NY, USA, 2008; pp. 348–402.
44. Willie, M.R.J.; Spangles, M.B. Application of electrical resistivity measurements to problem of fluid flow in porous media. *Am. Assoc. Pet. Geol. Bull.* **1952**, *36*, 359–403.
45. Del Rey, R.; Bertó, L.; Alba, J.; Arenas, J.P. Acoustic characterization of recycled textile materials used as core elements in noise barriers. *Noise Control Eng. J.* **2015**, *63*, 439–447. [CrossRef]
46. Glé, P.; Blinet, T.; Guigou-Carter, C. Acoustic performance prediction for building elements including biobased fibrous materials. In Proceedings of the Euronoise 2018, Crete, Greece, 27–31 May 2018; pp. 27–31.
47. Piégay, C.; Glé, P.; Gourdon, E.; Gourlay, E.; Marceau, S. Acoustical model of vegetal wools including two types of fibers. *Appl. Acoust.* **2018**, *129*, 36–46. [CrossRef]
48. Carosio, F.; Alongi, J. Flame Retardant Multilayered Coatings on Acrylic Fabrics Prepared by One-Step Deposition of Chitosan/Montmorillonite Complexes. *Fibers* **2018**, *6*, 36. [CrossRef]
49. No, H.K.; Meyers, S.P.; Prinyawiwatkul, W.; Xu, Z. Applications of chitosan for improvement of quality and shelf life of foods: A review. *J. Food Sci.* **2007**, *72*, 87–100. [CrossRef]
50. Shehu, Z.; Lamayi, D.W.; Sabo, M.A.; Shafiu, M.M. Synthesis, Characterization and Antibacterial Activity of Kaolin/Gum Arabic Nanocomposite on Escherichia Coli and Pseudomonas Aeruginosa. *Res. J. Nanosci. Eng.* **2018**, *2*, 23–29.

materials

MDPI

Article

Performance Characterization of Broad Band Sustainable Sound Absorbers Made of Almond Skins

Stefania Liuzzi *, Chiara Rubino, Pietro Stefanizzi and Francesco Martellotta

Department of Civil Engineering Sciences and Architecture, Polytechnic University of Bari, via Orabona 4, I-70125 Bari, Italy; chiara.rubino@poliba.it (C.R.); pietro.stefanizzi@poliba.it (P.S.); francesco.martellotta@poliba.it (F.M.)

* Correspondence: stefania.liuzzi@poliba.it

Received: 4 November 2020; Accepted: 28 November 2020; Published: 1 December 2020

check for updates

Abstract: In order to limit the environmental impact caused by the use of non-renewable resources, a growing research interest is currently being shown in the reuse of agricultural by-products as new raw materials for green building panels. Moreover, the European directives impose the goal of sustainability supporting the investigation of passive solutions for the reduction of energy consumption. Thus, the promotion of innovative building materials for the enhancement of acoustic and thermal insulation of the buildings is an important issue. The aim of the present research was to evaluate the physical, acoustical, and thermal performances of building panels produced by almond skin residues, derived from the industrial processing of almonds. In this paper different mix designs were investigated using polyvinyl acetate glue and gum Arabic solution as binders. Air-flow resistivity σ and normal incidence sound absorption coefficient α were measured by means of a standing wave tube. Thermal conductivity λ, thermal diffusivity α, volumetric heat capacity ρc were measured using a transient plane source device. Finally, water vapor permeability δ_p was experimentally determined using the dry cup method. Furthermore, a physical characterization of the specimens in terms of bulk density ρ_b and porosity η allowed to study the correlation existing between the binder and the aggregates and the consequent acoustical and hygrothermal behavior occurring on the different mix designs. The achieved results suggested the investigated materials comparable to the main products currently existing on the market.

Keywords: agro-waste; sound absorption; hygrothermal performances; sustainable materials; circular economy

1. Introduction

A stunning growth in the quantity and variety of solid wastes produced by agricultural, mining, industrial, and domestic activities has been caused by the rapid increase of the population, urbanization, and the rise of living standards due to the technological innovations [1]. Thus, use of conventional materials such as bricks, mortar, and cement implies a huge thermal and electric energy consumption caused by the production process with a negative environmental impact. In a circular economic system, the use of by-products for creating new raw materials adds value to the waste, transforming discarded elements into useful new goods [2].

As a consequence, one of the research challenges is to convert waste into new raw matter in order to produce new building components to increase sustainability in the construction sector. Nowadays, a large demand has been placed on agricultural by-products in the development of sustainable building materials. Several research works have studied the potential reuse of aggregates derived from agricultural waste, investigating their efficiency in terms of load-bearing, acoustical, and hygrothermal performances for new ecological materials [3–5]. Almond is a crop of major importance worldwide with the production in 2011 being approximately two million tons, according to the FAO [6]. A summary of recent trends in almond production in Italy and harvested area measured in hectares is shown in Figure 1. In Italy the almond industry has different productive configurations, changing according to the management and the mechanization process [7].

A significant increase of the area harvested and the almond production from 2013 to 2018 can be appreciated in Figure 1.

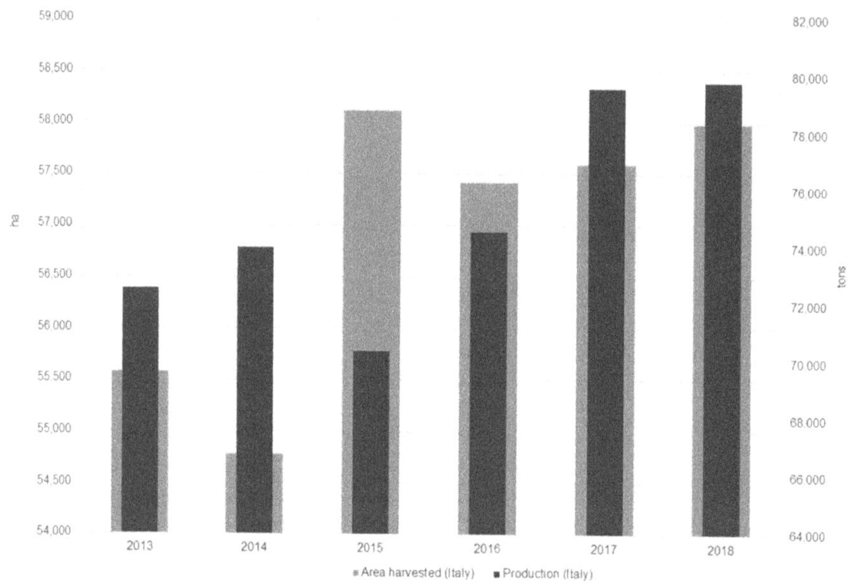

Figure 1. Summary of recent trends in almond production in Italy and harvested area measured in hectares (ha) obtained from elaboration of FAO data [6].

Liu et al. [8] investigated the multisector potential use of agricultural waste in the building construction field, finding different types of final applications: blocks, vegetable biomass for heating, multi-layer solutions, particles, coils, panels for the building envelope. The research on sound absorbing materials has become increasingly important throughout the last decades due to several possible fields of applications, from the aeronautics and road transportation industries to construction and buildings [9]. Currently, the main acoustic absorbing materials on the market have huge costs and represent a big issue in terms of final disposal due to their non-biodegradability [10]. Several researchers studied the potential use of natural components as sound absorbing materials [3,11–13]. Fouladi et al. [11] investigated the sound absorption coefficients of different panels made with coir, corn, sugar cane, and dry grass, finding that they are similar to the main common products now existing on the market. Asdrubali et al. [13] built an updated survey investigating the acoustical performances of sustainable materials developed with bio-based and recycled raw materials. They found a great availability on the market and a competitive price.

Martellotta et al. [12] studied the sound absorption properties of olive tree pruning waste with chitosan as a natural binder. They found very interesting performances, measuring absorption coefficients as high as 0.90 for the higher flow resistance sample at frequencies above 1 kHz. Mati-Baouche et al. [3] studied the thermal, acoustical, and mechanical properties of bio-based materials achieved by mixing crushed particles of sunflower stalks with chitosan, a natural binder.

The acoustical properties of a higher porosity composite were tested, achieving a good absorption coefficient.

Furthermore, in order to achieve high energy efficiency buildings, it is fundamental to control the hygric and thermal flux exchange between the internal and external environments. Thus, the use of high performance materials represents an efficient passive technique in lieu of active systems. Keynakli et al. [14] stated that thermal insulation is an essential parameter for building energy saving due to the high potential reduction of the rate of heat transfer. A material is considered as a thermal insulator when its thermal conductivity λ ($W \cdot m^{-1} \cdot K^{-1}$) is lower than 0.1 $W \cdot m^{-1} \cdot K^{-1}$ [15]. Korjenic et al. [16] developed a new insulating material from jute, flax, and hemp, demonstrating that bio-based materials are comparable to conventional ones for mechanical properties.

Luamkanchanaphan et al. [17] used a blend of narrow-leaved cattail fibers mixed with Methylene Diphenyl Diisocyanate (MDI), developing insulation boards. Thermal conductivity measurements ranged from 0.0438–0.0606 $W \cdot m^{-1} \cdot K^{-1}$ for panels with a density of 200–400 $kg \cdot m^{-3}$.

Therefore, in this framework, the development of new technologies based on renewable and natural sources was supported. The current growing interest in almond production derives from the fundamental nutrients constituting the basic fruit [18]. Almond is a fruit with high nutritional value; as a consequence, the industry has restricted the commercial relevance of almonds to the kernel. However, several authors have asserted that the other parts of the fruits (hull, shell, and skin) remain improperly underexplored [18,19].

Prgomet et al. [19] asserted that during the industrial process of removing the skin, two further by-products in addition to the shell and the hull are produced: blanched skin and blanch water (Figure 2). Thus, 70–85% of the whole almond fruit constitutes residues that currently are scarcely characterized and investigated. The main potential reuse of hulls and shells takes into account the physical and chemical properties of these by-products useful to the pharmaceutical and food sectors. The almond skin consists of carbohydrate polymer (cellulose) and an aromatic polymer (lignin).

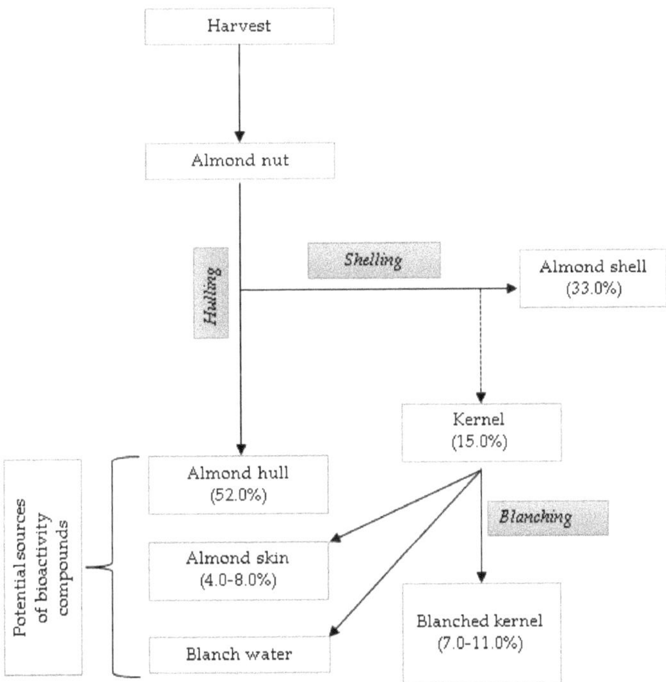

Figure 2. Process of obtaining almond by-products in the almond industry, adapted from Prgomet et al. [19].

Despite the existence of a great number of published papers reporting the main use of almond skin as powder for the production of biofuels like bio-ethanol [20], there are a few studies about the employment of this by-product as reinforcement filler in different thermoplastics [21–23]. Mankotia [23] found that in the PA6 plastic (Polyamide 6) the addition of the almond skin powder has positive effects on rheological, thermal, and wear properties. As a result, the valorization of such a by-product as new raw matter for building component production can represent an important way to add value to a low value product.

In order to use natural recycled fibers in the fabrication of porous building materials, a proper choice of the binder is essential. Adhesives play a fundamental role in a great range of areas from packaging to furniture to aerospace technology. The function of an adhesive is to bond separate components by a variety of mechanisms [24]. Most of the adhesives usually include a polymeric substance chemically bonded to the substrate. Heinrich [25] stated that the process and the boundary conditions of application are fundamental as the chemical composition because these factors influence the possible end uses. International standards require the reduction of formaldehyde [26], mainly used in wood furniture, in order to avoid harmful emissions during both the production process and the service life. A large number of researchers identified polyvinyl acetate glue as a suitable bonding material with good adhesive strength [27–29]. Ors et al. [28] and Khan et al. [29] focused on the improvement in adhesive strength of porous materials like wood and paper when using polyvinyl acetate glue.

In terms of sustainability, the use of adhesives based on bio and renewable materials can help to promote a circular economy by reducing the carbon footprint [25], even though their overall performance may show some limitations. Gum Arabic is a dried gummy existing on the current market, used mainly as a stabilizer and thickener in foods. However, several researchers described the good properties of gum Arabic, considering multiple industrial uses [30,31]. Because of its physical and mechanical properties, it represents a valid substitute for conventional binders mainly composed of formaldehyde. Rubino et al. [32] successfully used gum Arabic as a binder of wool waste fibers matrix for the production of sound absorbing materials.

In this scenario, the main aim of the present research was to develop new sustainable building composites, suitable as indoor covering materials with high acoustic and hygrothermal performances, using almond skin (AS) waste and two different types of binders: the polyvinyl acetate glue and gum Arabic. A comparison of their performances along with the properties of other building materials with similar bulk density highlighted that the AS materials are promising because of their equivalent or even better performance, compared with the main existing products currently available on the market.

2. Materials and Methods

2.1. Raw Materials

Almond skin wastes were used as bio-based aggregates and were provided by ALFRUS srl (Bari, Italy). Two different binders were selected: polyvinyl acetate (PVA) aqueous solution and a biodegradable gum Arabic (GA) aqueous solution (Figure 3).

PVA is a thermoplastic, formaldehyde-free resin available as a water-based emulsion achieved by the polymerization of vinyl acetate monomer; it has a density of 1009 kg·m^{-3}. PVA is a low cost, synthetic vinyl polymer with an invisible bond line, prepared by the polymerization of vinyl acetate monomer having the generic formula $(C_4H_6O_2)_n$. It is a type of thermoplastic that softens at 30–45 °C. Water emulsions of PVA are used to bind porous materials such as paper, wood, and clothes and to strengthen porous stone. A poor resistance to creep under load has been demonstrated. The clear film produced by the PVA has a great biodegradation resistance and a good resistance to weather, withstanding water, oil, grease, fire. It is mainly adopted as a nonstructural adhesive for projects at room temperature.

Gum Arabic is a complex polysaccharide in the form of natural gum derived from the hardened sap of the Acacia tree. It has a high molecular weight and it is water soluble.

Gum Arabic, a dried exudate from the Acacia Senegal tree, exists in several tropical and semitropical countries and is considered a biopolymer. The aqueous viscosity is strictly related to the origin of the products, the pH level, the electrolyte content, and the pretreatment. The addition of water to gum Arabic results in the reduction of the surface tension, achieving a solution of yellow–red color.

The aggregates were obtained during the manufacturing process of the hulls after peeling the almonds in high temperature water; thus, the thin layer of brown skin was removed and stored in external tanks. Before preparing the mix, the vegetable aggregates were oven-dried at 50 °C until they achieved constant mass ±0.5%, then cooled to environmental temperature in desiccators containing silica gel. The mixtures under study resulted from the heuristic search for the best combination of cohesion between the almond aggregates and the binders and the higher bulk porosity of the final specimens. The first mix, labeled AS0_GA, was based on almond skins used "as is," mixed with a GA water solution. The average width of the aggregates was 1 cm, the average length was 2 cm and the thickness was 0.5 mm. The second mix, labeled AS1_PVA, included ground almond skin fibers with an average grain size of 3 mm × 3 mm and 0.2 mm of thickness bonded with a PVA water solution.

Figure 3. Aggregates used. (On the left) almond residue in its true state used for AS0_GA, (on the right) almond residue in ground state used for AS1_PVA.

2.2. Sample Preparation

Many elements during the manufacturing process in the laboratory can influence the efficiency of the final samples: the initial water content of the fibers, the particle size distribution of the aggregates, the aggregates/binder ratio, and the type of the binder. Thus, several compositions were preliminarily tested in order to assess the best cohesion between the aggregates and the binders while keeping porosity as high as possible. All the samples were prepared under controlled environmental conditions to ensure that both materials and binders could interact in the best possible way and curing time could be standardized. A water solution was used to mix PVA glue and dissolve GA powder in a magnetic stirrer at normal temperature and relative humidity. The best quantities and ratios resulting from the preliminary tests are shown in Table 1.

Figure 4 presents the superficial morphology of the specimens. An optical microscope was used to assess the binding effect and to study the interaction between the Gum Arabic and polyvinyl acetate solutions and the aggregates.

The interface between the almond skin and the two different binders is highlighted. A superficial coat due to the binder can be noted on both types of the specimens. The rough surface morphology of the almond skin contributes to enhance the adhesion between the binder and the fiber in composite production, leading to stronger interfacial adhesion.

Table 1. Details of the different mix designs.

Solution	Almond Skin (A) (g)	Gum Arabic (B) (g)	Polyvinyl Acetate (B) (g)	Water (g)	B/A Ratio
AS0_GA	480	150	-	750	0.30
AS1_PVA	480	-	300	160	0.62

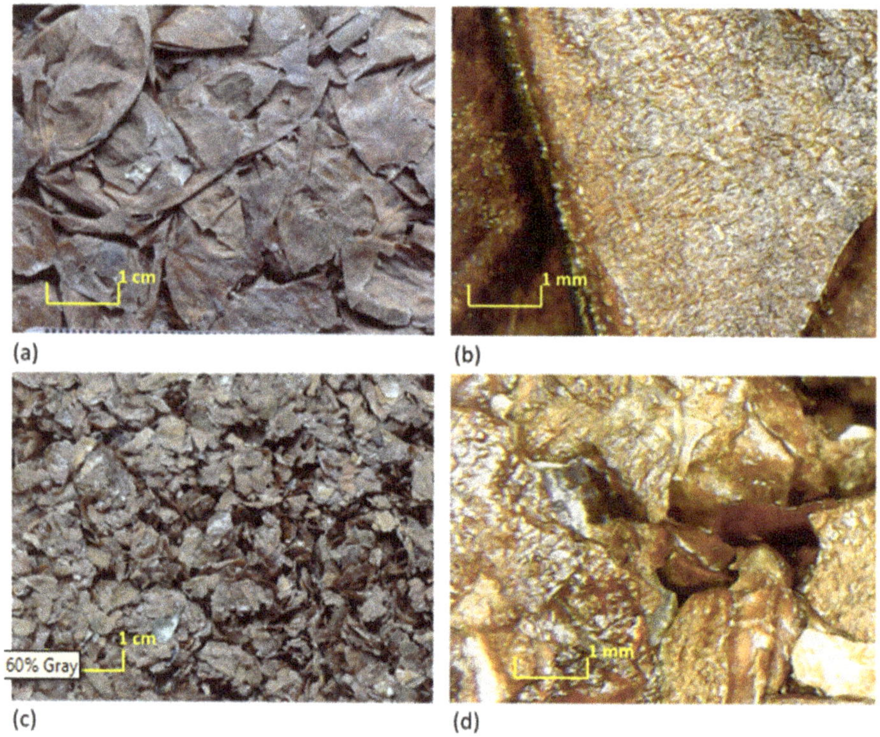

Figure 4. Magnified images of the analyzed samples: (**a**,**b**) AS0_GA; (**c**,**d**) AS1_PVA.

3. Methods

3.1. Physical Measurements

According to ISO 12570 [33] the bulk density ρ_b (kg·m^{-3}) of the specimens was calculated taking into account the size and the weight of the specimens. Each dimension was the average of three different measurements recorded by an electronic caliper (0.1 mm accuracy); and the weight was registered by an analytical balance (0.01 g accuracy). According to ASTM D4892 [34], the true density ρ_t (kg·m^{-3}) of the specimens was measured by an ULTRAPYC 1200-e helium gas pycnometer (Quantachrome, Boynton Beach, FL, USA) in order to determine the bulk porosity ε as follows:

$$\varepsilon = 1 - \frac{\rho_b}{\delta_p} \tag{1}$$

3.2. Thermal Measurements

The measurement of the dry-state thermal conductivity λ ($W{\cdot}m^{-1}{\cdot}K^{-1}$), the thermal diffusivity a ($10^{-6}{\cdot}m^2{\cdot}s^{-1}$), and the volumetric heat capacity ρc ($J{\cdot}m^{-3}{\cdot}K^{-1}$) were performed by an ISOMET 2104 (Applied Precision Ltd., Bratislava, Slovakia), a transient plane source device (Figure 5). The error of the measurement of these thermal parameters was estimated to be within ±4%, ±5%, and ±7%, respectively, according to Bouguerra et al. [35]. For each mix type, three representative samples of 10 cm diameter and 4 cm thickness were taken into account, and the values shown in Table 2 are the mean of three measurements. Before recording the measurements, the specimens were dried in a hot-air oven at 50 °C. Then, all the specimens were stabilized at 23 °C in desiccators containing silica gel.

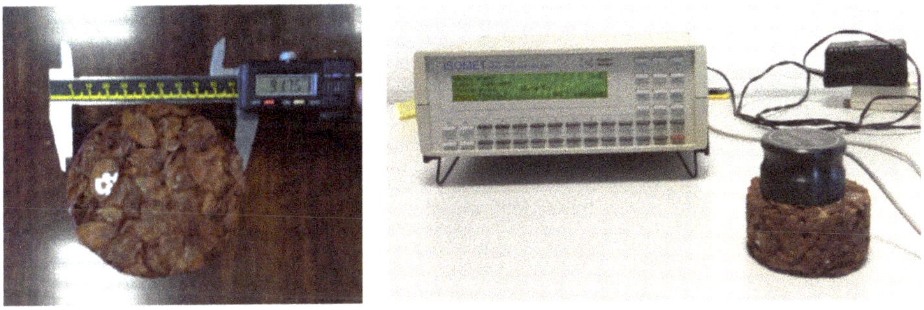

(a) (b)

Figure 5. Thermal measurement set up: (**a**) specimen dimensions measurement, (**b**) ISOMET 2104.

3.3. Hygric Measurements

The hygric parameters of the composites were determined by measuring the water vapor resistance coefficient μ (-) according to ISO 12572 [36] in a dry cup. Before starting the test, all the specimens were conditioned at 50 °C in a hot-air oven until achieving a change of mass less than 0.5%. Then, the samples were wax-sealed on the top of transparent vessels containing silica gel as desiccant (Figure 6a,b); an air space of 1.5 ± 0.5 cm was left between the desiccant and the sample. The assemblies were then placed in a temperature- and humidity-controlled environment chamber, Angelantoni DY340 (Angelantoni Test Technologies Srl, Massa Martana PG, Italy), set to 23 °C and 50% RH (Figure 6c).

The mass variations were calculated by recording daily mass using a Mettler PJ300 balance (±0.01 g accuracy, Mettler-Toledo GmbH, Greifensee, Switzerland) until achieving a constant mass loss for unit time. The water vapor resistance coefficient μ was then calculated from the experimental data as follows:

$$\mu = \frac{\delta_a}{\delta_p} \tag{2}$$

where $\delta_a = 1.94 \times 10^{-10}$ $kg{\cdot}m^{-1}{\cdot}s^{-1}{\cdot}Pa^{-1}$ is the water vapor permeability of air at 23 °C.

The water vapor permeability, δ_p, of the specimen was estimated from the following Equation:

$$\delta_p = \Lambda \cdot d \tag{3}$$

where d is the specimen thickness, Λ is the vapor permeance, calculated from Equation (4):

$$\Lambda = \frac{1}{\frac{A \cdot \Delta p}{\Delta G/\Delta \tau} - R_A} \tag{4}$$

The vapor pressure gradient Δp is the difference of partial vapor pressure between the air gap in the cup and the air in the climate chamber, A is the specimen area, R_A is the water vapor diffusion resistance of the air layer in the cup, and $\Delta G/\Delta\tau$ is the rate of change in mass.

(a) (b) (c)

Figure 6. Hygric measurements set up: (**a**,**b**) samples prepared in dry cups; (**c**) Angelantoni DY340 climate chamber.

3.4. Acoustic Measurements

Sound absorption measurements were performed by the transfer function method in order to determine the normal incidence sound absorption coefficient according to ISO 10534-2:1998 [37]. Two tubes with different diameters (10 cm, 4 cm) and a thickness of 5 mm were used for the test with the aim to consider the largest spectrum range (Figure 7). Thus, two groups of three specimens were tested, the first one with a 10 cm diameter and the second one with a 3 cm diameter. The tube with an internal diameter of 10 cm had a maximum measurable frequency of 2 kHz, and it used two different microphone distances (6 cm and 20 cm, respectively, yielding a low-frequency limit of 400 Hz and 50 Hz). The emitting end consisted of an 11 cm loudspeaker sealed into a wooden case and suitably isolated from the tube structure by an elastic and protective layer. For the second tube with a diameter of 4 cm the microphone distance was set to 3 cm, and the frequency covered a range between 200 Hz and 5 kHz. In both cases, the sample holder allowed different mounting conditions, close to the rigid termination and at distances that, in the present case, varied between 5 and 10 cm from the wall. All the results were processed by a MATLAB® (2018, Mathworks, Natick, MA, USA) graphical user interface generating a 5 s linear sweep from 70 Hz to 3 kHz, when used in combination with the largest tube, and from 500 Hz to 5 kHz when considering the smallest tube. The performance of the previously described device was compared using the same material (a 5 cm polyester fiber sample) as that of the BSWA SW 260 two-microphone impedance tube, showing that one-third octave band values differed by less than 5% over the entire overlapping spectrum.

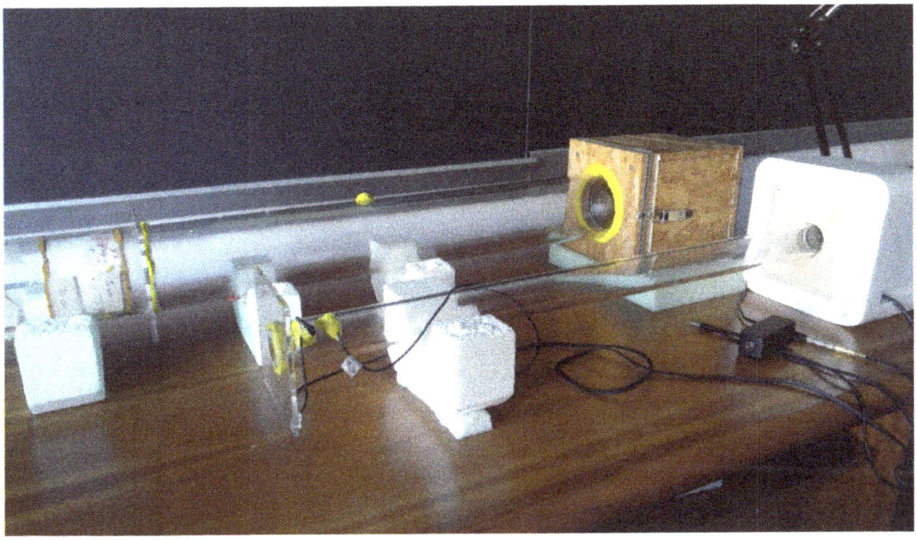

Figure 7. Photograph of the standing wave tubes used for sound absorption at low and medium frequencies (top), and sound absorption at high frequencies and flow resistance measurements (bottom).

3.5. Non-Acoustic Measurements

Non-acoustic parameters usually include the descriptors of the porous structure of the material. Their number and type are strongly dependent on the model used to describe the sound absorbing layer. Several motionless skeleton (or "equivalent fluid") models were available in the literature, spanning from the simplest ones, based on macroscopic empirical models and developed by Delany and Bazley [38] and later improved by Miki [39] (only depending on air-flow resistivity), to the most refined semi-phenomenological and microstructural models. Among them, it was possible to account for the early models reviewed by Attenborough [40], the microstructural model the same author proposed [41], basing the prediction of the acoustical characteristics of rigid fibrous absorbents on five parameters (porosity, flow resistivity, tortuosity, steady flow shape factor, and dynamic shape factor), and possibly others. However, it was the semi-phenomenological model developed by Johnson et al. [42] and subsequently refined by Allard and Champoux [43] that, by using only five non-acoustic parameters (including air-flow resistance σ, porosity ε, tortuosity k_s, and viscous Λ and thermal Λ' characteristic lengths), offered a good balance between ease of use and prediction accuracy. Further improvements to the above formulation were given by Lafarge et al. [44], including one more parameter named static thermal permeability $k_{0'}$, and Pride at al. [45], including the static viscous tortuosity α_0 and the static thermal tortuosity $\alpha_{0'}$.

Considering that only a small subset of the previously mentioned parameters can be directly measured, and that the others need to be estimated according to the procedures described below, the Johnson–Champoux–Allard (JCA) model [42,43] was preferred as the reference one to help in interpreting the results. This model is one of the most frequently used in the literature and is implemented in several vibro-acoustic tools. In addition, the additional parameters required by the more complex models rely on a limited literature background, without considering that the variations they introduce are typically small and refer only to low frequencies. However, just for comparison purposes, results obtained with the Johnson–Champoux–Allard–Lafarge (JCAL) model [44] were also shown for the base cases. Equations used to derive the effective density and the effective bulk modulus according to the proposed models, and mostly derived from Allard and Atalla [46], are described in detail in Appendix A.

Thus, in light of the previously mentioned choice, the non-acoustic parameters considered in this paper were: air-flow resistance σ, porosity ε, tortuosity k_s, viscous Λ and thermal Λ' characteristic lengths, and static thermal permeability $k_{0'}$. Among them, porosity was measured directly as described in Section 3.1., as well as air flow resistivity, which was measured according to the acoustic method proposed by Ingard and Dear [47]. According to this setup the sample was located in between two 85 cm long methacrylate tubes (with a 4 cm inner diameter), having a rigid termination on one side and a loudspeaker (Visaton FRS 5, Visaton, Haan, Germany) on the other. The loudspeaker had a flat response between 150 Hz and 20 kHz and was fed by an exponential sine sweep. The pressure drop through the sample was determined by means of two calibrated microphones (Core Sound, Teaneck, NJ, USA) located close to the sample and to the rigid termination allowed to calculate.

As it was impossible to directly measure the remaining parameters, they were estimated. An inverse method [48] was used to estimate the missing ones. Taking advantage of measured absorption coefficients, porosity, and air-flow resistivity, the values of the missing physical properties (i.e., tortuosity, viscous and thermal lengths ratio, and shape factor) were determined by means of a brute-force algorithm developed in MATLAB® to find the set of parameters that allowed the best match between measurements and predictions. Measured parameters were also allowed to vary within their uncertainty ranges, while the properties that were not measurable varied over the entire range of possible values. In particular, tortuosity values between 1 and 6 were explored, while for the shape factor (on which the viscous characteristic length is directly dependent), values between 0.3 and 3.3 were explored. Finally, the ratio of the characteristic lengths varied between 1 and 10, to satisfy the condition that $\Lambda \leq \Lambda'$. The cost function minimized by the search algorithm was the mean absolute error between measured and predicted one-third octave absorption coefficients in the range from 100 Hz to 3150 Hz.

4. Results and Discussion

4.1. Thermal and Hygric Characterization

Table 2 shows the results of the thermal properties and hygric measurements executed according to ISO 12570 in a dry cup for the two different blends. In general, a strict correlation between the hygrothermal parameters and the bulk density exists. The reduction of the bulk density corresponds to an increase of porosity and, as a consequence, an enhancement of the thermal parameters and hygric properties. For each mix three representative specimens were evaluated and the final value of the parameters, reported in Table 2, was calculated as the mean of the three measurements. The standard uncertainties were calculated in accordance with JCGM 100:2008 [49]. Both the experimental materials can be considered good thermal insulators as the thermal conductivity was less than 0.1 $W \cdot m^{-1} \cdot K^{-1}$ in compliance with Al-Homoud [15]. As shown in Figure 8, the thermal conductivity obtained for the AS0_GA and AS1_PVA samples was lower than 0.2 $W \cdot m^{-1} \cdot K^{-1}$; the recorded values were in both cases of the same order as many conventional insulators such as EPS and mineral wool [50]. Furthermore, a good similarity of thermal conductivity could be appreciated in different literature cases referring to the thermal properties of agro-waste materials. Considering the bulk density, the AS1_PVA sample had the thermal conductivity comparable to a rice husk panel [51]. The thermal properties of the AS0_GA specimen were included in the range of typha and date palm wood [17,52–54].

In theory, depending on the degree of porosity, the vapor permeability diffusion resistance factor is expected to be lower when a reduction in bulk density occurs. As a matter of fact, AS0_GA had a μ-value of 13.036 lower than the water vapor diffusion resistance factor of AS1_PVA.

Comparing the experimental specimens with other sustainable materials available in the literature, it was deduced that the value of the AS material was in accordance with bamboo particleboards, with a μ-value ranging from 9.2 and 12.8 [55].

Table 2. Summary of hygrothermal properties for each mix. Values in brackets represent standard deviations of measured values.

Mix Code	Bulk Density	True Density	Thermal Conductivity	Thermal Diffusivity	Volumetric Heat Capacity	Water Vapor Resistance
	ρ_b (kg·m^{-3})	ρ_t (kg·m^{-3})	λ (W·m^{-1}·K^{-1})	α (10^{-6} m^2·s^{-1})	ρc (10^6 J·m^{-3}·K^{-1})	μ (-)
AS0_GA	234.65 (13.99)	1255.29 (19.88)	0.074 (0.0035)	0.180 (0.0276)	0.416 (0.0533)	13.0 (0.2)
AS1_PVA	373.10 (6.20)	1162.58 (7.29)	0.082 (0.0080)	0.219 (0.0342)	0.384 (0.0745)	14.9 (1.4)

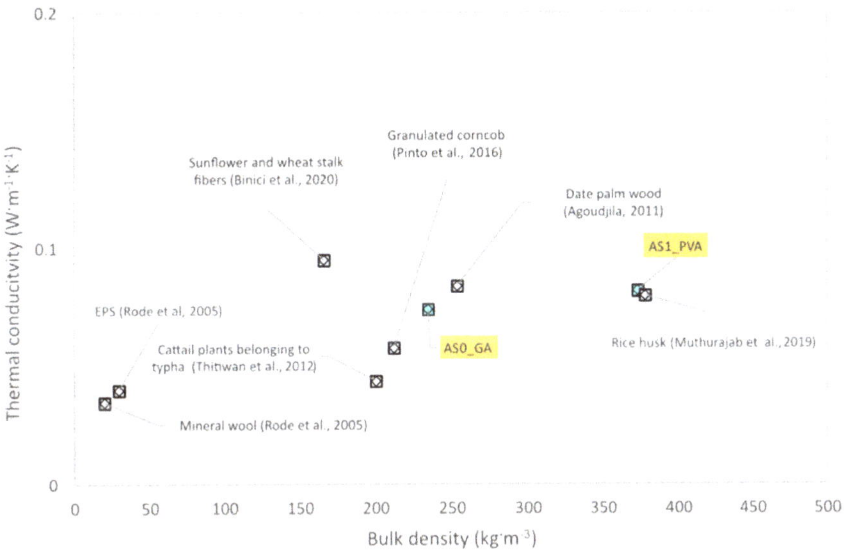

Figure 8. Comparison of thermal conductivity between AS0_GA, AS1_PVA tested samples and different insulators, taken from the literature.

4.2. Acoustic and Non-Acoustic Parameters

The results for normal incidence sound absorption coefficients are shown in Figure 9. The two samples showed nearly the same behavior with two evident peaks spaced out by a drop, which is typical of thick porous materials, although the frequency where the first peak appeared was shifted toward higher frequencies for the AS1_PVA sample. The low frequencies behavior, up to 160 Hz, was rather similar for both cases. The AS0_GA sample sharply increased up to 500 Hz, where the first peak appeared, with α rising up to almost 0.8. Then, a drop appeared with a minimum α value of 0.6, followed by a new peak at 3150 Hz. The AS1_PVA sample showed absorption coefficients lower than the AS0_GA ones up to 630 Hz, where the first peak appeared, with α rising to 0.9; while the AS0_GA curve dropped with α of 0.7. Then, a drop appeared with a minimum absorption value around 0.6, followed by a new peak at 3150 Hz, almost overlapping the second peak of the AS0_GA sample. For porous materials, normal incidence sound absorption is characterized by peaks and valleys, with peaks appearing when particle velocity inside the medium is at a maximum. Assuming that the backing surface is rigid, this first peak takes place at a frequency whose quarter wavelength corresponds to the thickness of the sample and all the others at odd multiples of that frequency [46]. In the present case, for a 5 cm thick panel, the first peak should be at 1.7 kHz, assuming the speed of sound inside the porous medium is the same as that of the air. However, as the first peak was shifted toward lower frequencies in both cases, it was to be expected that, as shown by many authors [3,12,56], due to the more complex and tortuous pore structure, the actual speed of sound was significantly lower. Finally, the subsequent peaks expected at odd multiples of the first peak frequency appeared somewhat smoothed into a single high-frequency peak.

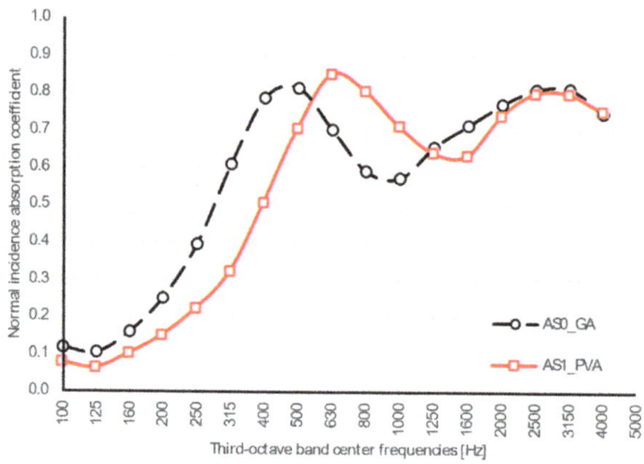

Figure 9. Plot of normal incidence absorption coefficients measured in one-third octave bands for samples with polyvinyl acetate (AS1_PVA) and gum Arabic (AS0_GA) specimens.

Taking into account Figure 10, a comparison with same-thickness samples made of olive trees [12] with bulk density (between 220 and 240 kg·m^{-3}) comparable with that of the AS0_GA specimen but with much lower flow resistivity (4.2 kN·s·m^{-4}), and with recycled textile materials with similar resistivity (14.4 kN·s·m^{-4}) but different density (93 kg·m^{-3}), pointed out quite different behaviors, with the almond skin sample showing the first peak significantly shifted toward lower frequencies. Similarity in density, as expected, had very little influence on the results, but even similarity in flow resistivity proved not to be enough to ensure a similar behavior; this suggested that the different first peak positions could be attributed to the major role played by porosity and the morphology of the internal cavities of the specimens.

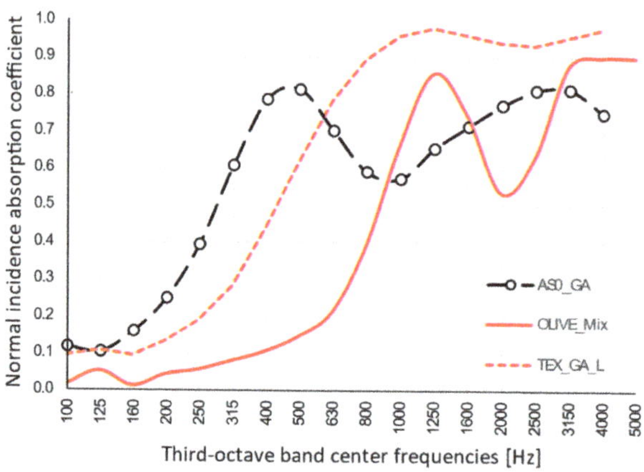

Figure 10. Comparison of measured results for sample with gum Arabic binder (AS0_GA) and data from literature referring to samples made of olive trees (OLIVE_MIX) [12] having same density or specimens with same flow resistivity (TEX_GA_L) [32].

The application of the phenomenological model defined by Johnson–Champoux–Allard [42,43] and the subsequent indirect determination of the characterizing parameters returned a very good agreement between measured and predicted absorption coefficients, with peaks appearing at the same frequencies, although with slightly different absolute values (Figure 11). The JCAL model [44] was also applied in this case for comparison purposes and apparently provided no significant improvement, compared to the JCA model. In fact, the minimum value of the cost function was slightly higher for the JCAL model than for the simpler one. Predicted speed of sound inside the porous structure was frequency dependent, and its value at the first absorption peak was about 120 m/s for sample AS0_GA and 150 m/s for sample AS1_PVA. The analysis of the parameters resulting from the indirect measure (Table 3) suggested that the multiple layers of thin almond skins originated a very complex channel pattern that resulted in high tortuosity, spanning between 4 and 5.5 values. Such results appear significantly higher than the values typically found in the literature with reference to crushed materials [3,57], which are typically around 3. However, it is possible to find even higher tortuosity values when dealing with open porous asphalt [58]. Following this latter statement, even if high values of tortuosity, according to most empirical formulas [59], are typically associated with very low porosity of the medium, it should be considered that the relationship between porosity and tortuosity is strongly dependent on the shape of the grains that compose the solid matrix, as demonstrated by Turo and Umnova [60]. As reported by Sarradj et al. [58] a general relation between the two variables takes the form of $k_s = \varepsilon^{-L/(1-L)}$, where L is a factor depending on the shape of the particles, varying between 0 when particles are needles parallel to flow, and 1 when particles are disks perpendicular to flow. In the latter case, tortuosity may become very high even when porosity is high. Considering that almond skins, particularly those used "as is," tend to distribute according to layers mostly perpendicular to the faces exposed to sound (Figure 12), the observed values of tortuosity appear perfectly consistent with that structure. In particular, for AS0_GA the whole skins tended to create many more connected pores and consequently had a lower density and higher porosity. Conversely, AS1_PVA samples had the same layered structure (with some "vertical" inclusions here and there (Figure 12d)), but were much more compact and dense.

Once the origin of the low-frequency shift was clarified, as this may be a desirable feature to obtain in sound absorbing materials that extend their effectiveness range to low frequencies (which are always more difficult to treat while keeping overall thickness within reasonable limits), a further experiment was carried out in order to check consistency of the hypothesis and investigate potential use of the panels under different mounting conditions, which might extend the frequency range in which the panel are efficient.

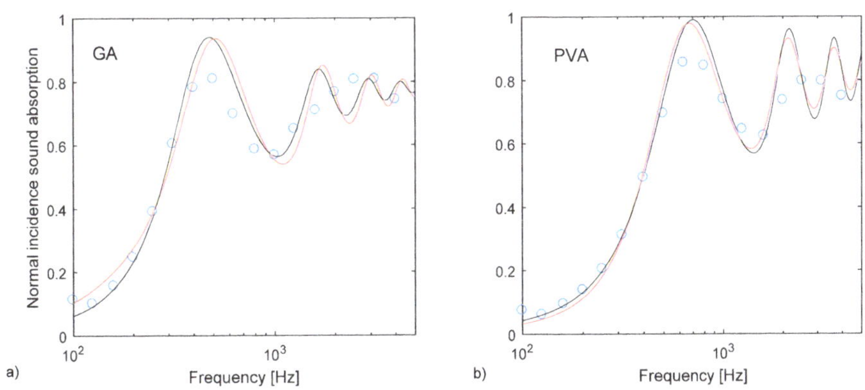

Figure 11. Comparison of measured and best-fit values normal incidence sound absorption coefficients resulting from the application of the JCA model (black curve) and the JCAL model (red curve): (**a**) AS0_GA sample; (**b**) AS1_PVA sample.

Table 3. Summary of non-acoustical parameters measured (underlined) or indirectly derived from the application of the JCA (italicized) and JCAL models. Values in brackets represent standard deviations.

Sample Code	Model	Bulk Porosity ε (-)	Air-flow Resistivity σ ($kN{\cdot}s{\cdot}m^{-4}$)	Tortuosity (-)	Shape Factor (-)	Ratio of Characteristic Dimensions (-)	Static Thermal Permeability (m^2)
AS0_GA	JCA	0.81	13.377	*5.50*	*3.00*	*1.50*	
	JCAL	(0.003)	(1.164)	5.40	3.00	2.00	3.1×10^{-9}
AS1_PVA	JCA	0.68	23.371	*4.40*	*1.36*	*1.95*	
	JCAL	(0.002)	(1.791)	5.00	2.36	1.80	2.1×10^{-9}

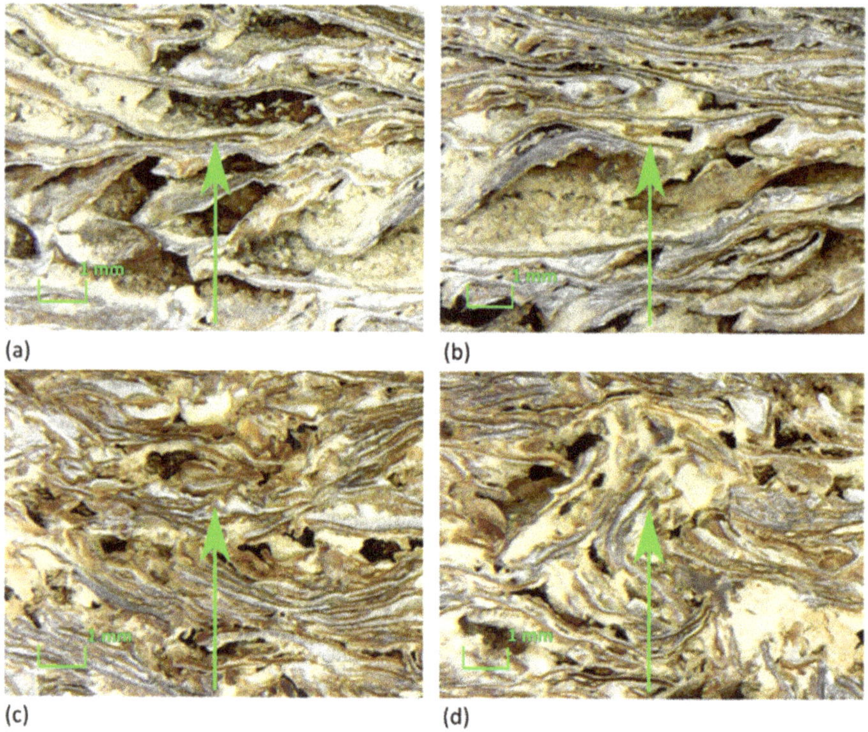

Figure 12. Magnified images of the cross section of the analyzed samples: (**a**,**b**) AS0_GA; (**c**,**d**) AS1_PVA. In all the cases sound propagation takes place along a vertical direction (green arrow).

In fact, if the previous interpretation of the shift of the first peak toward low frequency was correct, pointing out the role played by tortuosity and morphology of the porous matrix on the absorption coefficient, mounting the sample at a different distance from the rigid termination should determine a further movement of the peak toward even lower frequencies. Results shown in Figure 13 confirmed that the first peak was shifted exactly according to the to the JCA model predictions (particularly for AS1_PVA samples), although in the high-frequency range the fluctuations between peaks and valleys were less evident than they were in the prediction model.

These results suggested that panels made of almond skins, even in the case where they are made by simply mixing them with some sort of binder, may provide a noticeable absorption even in the low-frequency range despite the use of a 5 cm thick panel. The addition of an extra 10 cm air gap (which could be easily obtained in false ceiling application) further contributed to extend the absorption down to 140 Hz for AS0, where a sound absorption higher than 0.7 was found, but in the range between 330 Hz and about 1 kHz values fell again below 0.7. Similar behavior was found for AS1 for which the frequency where the absorption coefficient first reached 0.7 was 160 Hz, while the dip extended from 500 Hz to 1250 Hz. It was interesting to observe that under diffuse field conditions (Figure 14), predicted by calculating a spatially weighted average of the absorption coefficients resulting from different incident angles [46], the dip between the first and second peak appeared significantly attenuated, and the frequency dependent response was generally much smoother. In this case, even using the 5 cm panel without air gap a sound absorption coefficient higher than 0.7 could be obtained from 300 Hz on for AS0_GA and from 500 Hz on for AS1_PVA. Using the 10 cm air gap allowed to push down the frequency where sound absorption reached 0.70 at 170 Hz for the AS0_GA panel and at about 200 Hz for the AS1_PVA panel, thus suggesting that the panels under investigation might be conveniently used as broad band sound absorbers.

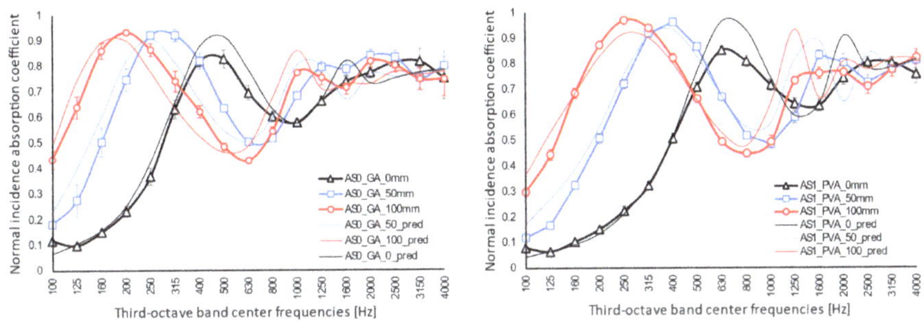

(**a**) (**b**)

Figure 13. Comparison of measured and predicted normal incidence absorption coefficients resulting from different mounting distances (0, 50, and 100 mm) from hard surface for: (**a**) AS0_GA samples, and (**b**) AS1_PVA samples. Predicted values are based on the application of the JCA model with parameters optimized only for the sample mounted on a rigid surface.

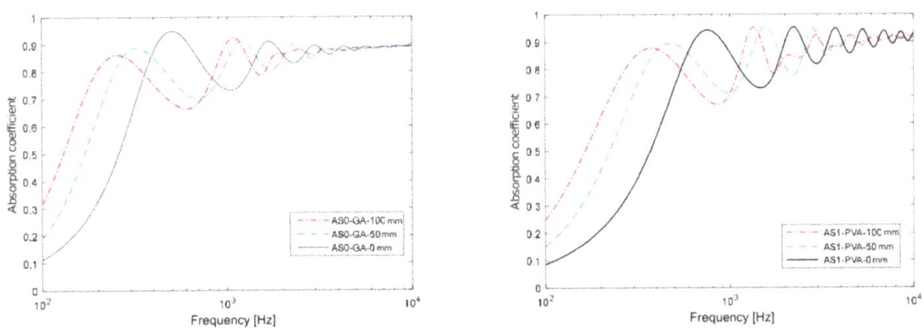

(**a**) (**b**)

Figure 14. Comparison of diffuse field sound absorption coefficients as a function of frequency predicted, using the JCA model with reference to the sample mounted on a rigid surface and at increasing distances from it. (**a**) Values for sample AS0 with GA binder; (**b**) Values for sample AS1 with PVA binder.

5. Conclusions

This study focused on a possible way to recycle the almond skin, an agro-waste derived from the almond production process, transforming it into new aggregates used for the production of almond skin materials. Different kinds of tests were performed in order to characterize the hygric, thermal, and acoustical properties of the materials. The results of the experimental campaign allowed to draw the following conclusions.

In general, a good adhesion between the binder and the AS residues was observed, taking into account the morphology and the cross section of the specimens (Figures 4 and 12); the aggregates were well coated and bound together. In particular, for AS0_GA, the whole skins tended to create many more connected pores and consequently had a lower density and higher porosity. Conversely, AS1_PVA samples had the same layered structure (with some "vertical" inclusions here and there) but were much more compact and dense. In fact, although no significant variation of the true density could be appreciated (Table 2), a large variation in the bulk density and bulk porosity ε was observed. As expected, it was found that lowering the fiber size caused an increase in the bulk density due to less volume being occupied by the almond skin residues at lower sizes.

Considering the thermal properties, it was found that, due to the thermal conductivity less than 0.1 $W{\cdot}m^{-1}{\cdot}K^{-1}$, both AS0_GA and AS1_PVA samples behaved as good insulators. In addition, the thermal conductivity of both the samples was comparable to the thermal performance of other agro-waste materials and of widespread insulators like EPS and mineral wool. The hygric properties of the experimental specimens, as well, were in the range of bamboo particleboards with a μ-value between 9.2 and 12.8.

From the acoustic point of view, the results of the normal incidence sound absorption coefficients measurement showed that both the samples outperformed the typical behavior of thick porous materials. In fact, for typical porous materials, peak location was strongly dependent on the thickness of the sample. In this case it was expected at 1.7 kHz, significantly above the actual observed location. This result was explained in terms of a more complex and tortuous morphology of the pore structure of the specimens.

A very good agreement between measured and predicted absorption coefficients, with peaks appearing at the same frequencies, was achieved using the phenomenological model defined by Johnson–Champoux–Allard [38]. Some parameters were indirectly measured and a tortuosity ranging between 4 and 5.5 was found, in good agreement with the complex channel pattern visible in the cross-section images (Figure 12). In particular, for the AS0_GA specimen it was observed that layers mostly perpendicular to the faces (and to sound propagation) were created. This justified the observed values of tortuosity, perfectly in accordance with that structure.

The previously described experimental study broadly supported the use of almond skin waste to manufacture innovative and sustainable building materials. The physical, hygrothermal, and acoustic properties demonstrated the high potential for such building components.

Further investigations are underway to assess the mechanical performances and the fire protection properties, moreover, evaluating the suitability of other sustainable and eco-friendly binders to be used with the almond skin waste. Furthermore, an LCA analysis and a techno-economic feasibility study for the induction of a building panel on market are necessary and they will be carried out to support future potential commercial applications.

Author Contributions: Conceptualization, S.L.; Methodology, S.L., F.M., C.R. and P.S.; Investigation, S.L. and C.R.; Formal analysis, S.L., F.M., C.R. and P.S.; Data curation, S.L., C.R. and F.M.; Writing—original draft preparation, S.L. Writing—review and editing, S.L., F.M., C.R. and P.S.; Visualization, S.L. and F.M.; Software, F.M.; Supervision, P.S. and F.M.; Funding acquisition, P.S. and F.M. All authors have read and agreed to the published version of the manuscript.

Funding: The research was carried out within the framework of the Italian PRIN ("Progetto di Ricerca di Rilevante Interesse Nazionale) Project "SUSTAIN/ABLE—SimultaneoUs STructural and energetIc reNovAtion of BuiLdings through innovativE solutions", ERC Sector PE8, ID 20174RTL7W_007.

Acknowledgments: Authors are grateful to ALFRUS srl (Bari, Italy) for providing almond skins used to make the investigated samples.

Conflicts of Interest: The authors declare no conflict of interest.

Appendix A

In order to calculate the sound absorption coefficient according to the JCA model, the dynamic (or effective) density and bulk modulus are requested. The first is based on the work by Johnson et al. [42], where visco-inertial dissipative effects inside the porous media are described, and is given by:

$$\rho_e = k_s \rho_0 \left[1 + \frac{\sigma \varepsilon}{j \omega \rho_0 k_s} \sqrt{1 + j \frac{4 k_s^2 \eta \rho_0 \omega}{\sigma^2 \Lambda^2 \varepsilon^2}} \right] \tag{A1}$$

where ρ_0 is the static density, k_s is the tortuosity, σ is the flow resistivity, ε is the porosity, ω is the angular frequency, η is the dynamic viscosity, Λ is the viscous characteristic length and j is the complex unit.

The work by Champoux and Allard [43] is used to describe the thermal dissipative effects and calculate the dynamic bulk modulus according to the formula:

$$K_e = \frac{\gamma P_0}{\gamma - (\gamma - 1) / \left(1 + \frac{8 \eta}{j \Lambda'^2 N_p \rho_0 \omega} \sqrt{1 + j \frac{\Lambda'^2 N_p \rho_0 \omega}{16 \eta}} \right)} \tag{A2}$$

where γ is the ratio of the specific heat capacities, P_0 is the atmospheric pressure, N_p is the Prandtl Number, and Λ' is the thermal characteristic length.

Finally, according to Lafarge et al. [44], the dynamic bulk modulus is modified to account for low-frequency thermal effects, yielding the following formula:

$$K_e = \frac{\gamma P_0}{\gamma - (\gamma - 1) / \left(1 + \frac{\varepsilon \eta}{j k_0 N_p \rho_0 \omega} \sqrt{1 + j \frac{4 k_0^2 N_p \rho_0 \omega}{\eta \Lambda'^2 \varepsilon^2}} \right)} \tag{A3}$$

where k_0 is the static thermal permeability.

Once the dynamic density and bulk modulus have been determined it is possible to calculate the characteristic impedance z_c as

$$z_c = \sqrt{K_e \rho_e} \tag{A4}$$

and the propagation wave number as

$$k = \omega \sqrt{\frac{\rho_e}{K_e}} \tag{A5}$$

so that the absorption coefficient may be calculated according to the usual formulas, as well as other important properties like the actual dynamic speed (frequency dependent) that is given by ω/k.

References

1. Pappu, A.; Mohini, S.; Asolekar, S.R. Solid wastes generation in India and their recycling potential in building materials. *Build. Environ.* **2007**, *42*, 2311–2320. [CrossRef]
2. Rubino, C.; Bonet Aracil, M.; Liuzzi, S.; Stefanizzi, P.; Martellotta, F. Wool waste used as sustainable nonwoven for building applications. *J. Clean. Prod.* **2021**, *278*, 123905. [CrossRef]
3. Mati-Baouche, N.; De Baynast, H.; Michaud, P.; Dupont, T.; Leclaire, P. Sound absorption properties of a sunflower composite made from crushed stem particles and from chitosan bio-binder. *Appl. Acoust.* **2016**, *111*, 179–187. [CrossRef]

4. Liuzzi, S.; Rubino, C.; Martellotta, F.; Stefanizzi, P.; Casavola, C.; Pappalettera, G. Characterization of biomass-based materials for building applications: The case of straw and olive tree waste. *J. Ind. Crop. Pro.* **2020**, *147*, 11222. [CrossRef]
5. Madurwar, M.V.; Ralegaonkar, R.V.; Sachin, A.M. Application of agro-waste for sustainable construction materials: A review. *Constr. Build. Mater.* **2013**, *38*, 872–878. [CrossRef]
6. Available online: http://www.fao.org/home/en/ (accessed on 20 October 2020).
7. Sottile, F.; Massaglia, S.; Peano, C. Ecological and Economic Indicators for the Evaluation of Almond (*Prunus dulcis* L.) Orchard Renewal in Sicily. *Agriculture* **2020**, *10*, 301.
8. Liu, L.F.; Li, H.Q.; Lazzaretto, A.; Manente, G.; Tong, C.Y.; Liud, Q.B.; Ping, L.N. The development history and prospects of biomass-based insulation materials for buildings. *Renew. Sustain. Energy Rev.* **2017**, *69*, 912–932. [CrossRef]
9. Da Silva, A.R.; Mareze, P.; Brandão, E. Prediction of sound absorption in rigid porous 348 media with the lattice Boltzmann method. *J. Phys. A Math. Theor.* **2016**, *49*, 6.
10. De Carvalho, P.S.; Macklini, D.N.; Cantorski da Rosa, L. Development of an acoustic absorbing material based on sunflower residue following the cleaner production techniques. *J. Clean. Prod.* **2020**, *270*, 122478. [CrossRef]
11. Fouladi, M.H.; Nassir, M.H.; Ghassem, M.; Shamel, M.; Yeng Peng, S.; Wen, S.Y.; Xin, P.Z.; Nor Mohd, M.J. Utilizing Malaysian natural fibers as sound absorber. In *Modeling and Measurement. Methods for Acoustic Waves and for Acoustic Microdevices*; Beghi, M.G., Ed.; IntechOpen Limited: London, UK, 2013; pp. 161–170. ISBN 978-953-51-1189-4.
12. Martellotta, F.; Cannavale, A.; De Matteis, V.; Ayr, U. Sustainable sound absorbers obtained from olive pruning wastes and chitosan binder. *Appl. Acoust.* **2018**, *141*, 71–78. [CrossRef]
13. Asdrubali, F.; Schiavoni, S.; Horoshenkov, K.V. A review of sustainable materials for acoustic applications. *Build. Acoust.* **2012**, *19*, 283–312. [CrossRef]
14. Kaynakli, O. A review of the economical and optimum thermal insulation thickness for building applications. *Renew. Sust. Energ. Rev.* **2012**, *16*, 415–425. [CrossRef]
15. Al-Homoud, M.S. Performance characteristics and practical applications of common building thermal insulation materials. *Build. Environ.* **2005**, *40*, 353–366. [CrossRef]
16. Korjenic, A.; Petránek, V.; Zach, J.; Hroudová, J. Development and performance evaluation of natural thermal-insulation materials composed of renewable resources. *Energ. Buildings* **2011**, *43*, 2518–2523. [CrossRef]
17. Luamkanchanaphana, T.; Chotikaprakhan, S.; Jarusombati, S. A Study of Physical, Mechanical and Thermal Properties for Thermal Insulation from Narrow-leaved Cattail Fibers. *APCBEE Procedia* **2012**, *1*, 46–52. [CrossRef]
18. Özcan, M.M.; Ünver, A.; Erkan, E.; Arslan, D. Characteristics of some almond kernel and oils. *Sci. Hortic.* **2011**, *127*, 330–333.
19. Prgomet, I.; Berta Gonçalves, P.; Perles, R.D.; Seva, N.P.; Barros, A.I. Valorization Challenges to Almond Residues: Phytochemical Composition and Functional Application. *Molecules* **2017**, *22*, 1774. [CrossRef]
20. Sanjay, M.R.; Madhu, P.; Jawaid, M.; Senthamaraikannan, P.; Senthil, S.; Pradeep, S. Characterization and properties of natural fiber polymer composites: A comprehensive review. *J. Cleaner Prod.* **2018**, *172*, 566–581. [CrossRef]
21. Kumar, S.; Singh, R.; Singh, T.P.; Batish, A. PLA composite matrix as functional prototypes for four dimensional applications. In *Reference Module in Materials Science and Materials Engineering*, 2nd ed.; Elsevier: Amsterdam, The Netherlands, 2019; pp. 1–12, ISBN 978-0-12-803581-8.
22. García, A.V.; Santonja, M.R.; Sanahuja, A.B.; Garrigós, C. Characterization and degradation characteristics of poly (e -caprolactone) -based composites reinforced with almond skin residues. *Polym. Degrad. Stab.* **2014**, *108*, 269–279. [CrossRef]
23. Mankotia, K.; Shing, I.; Shing, R. On effect of almond skin powder waste reinforcement in PA6: Rheological, thermal and wear properties. *Mater. Today Proc.* **2020**, *33*, 1546–1551. [CrossRef]
24. Müller, B.; Rath, W. *Formulierung von Kleb- und Dichtstoffen: Das kompetente Lehrbuch für Studium und Praxis*; Vincentz Network GmbH & Co KG: Hannover, Germany, 2004; ISBN 978-3866306059.
25. Heinrich, L.A. Future opportunities for bio-based adhesives—advantages beyond renewability (Critical Review). *Green Chem.* **2019**, *21*, 1866–1888. [CrossRef]

26. Formaldehyde Emission. *Standards for Composite Wood Products*; US Government, Environmental Protection Agency (EPA): Washington, DC, USA, 2017; pp. 89674–89743. Doc. Number 2016–2798.
27. Mancini, E.; Antonelli, M.G.; Zobel Beomonte, P.; Sasso, M. Characterization and analytical parametrization of composite in cellulose fibre and PVA matrix. *Compos. Part B Eng.* **2019**, *172*, 496–505. [CrossRef]
28. Ors, Y.; Atar, M.; Ozçïfçï, A. Bonding strength of poly(vinyl acetate)-based adhesives in some woods materials treated with impregnation. *J. Appl. Polym. Sci.* **2000**, *76*, 1472–1479. [CrossRef]
29. Khan, U.; May, P.; Porwal, H. Coleman J. Improved adhesive strength and toughness of Polyvinyl Acetate Glue on addition of small quantities of graphene. *ACS Appl. Mater. Interfaces* **2013**, *5*, 1423–1428. [CrossRef] [PubMed]
30. Parija, S.; Misra, M.; Mohanty, A.K. Studies of natural gum adhesive extracts: An overview. *J. Macromol. Sci., Part C Polym. Rev.* **2001**, *41*, 175–197. [CrossRef]
31. Meer, W. Gum Arabic. Chapter 8. In *Handbook of Water Soluble Gums and Resins*; Davidson, R.L., Ed.; McGraw Hill: New York, NY, USA, 1980; ISBN 978-0070154711.
32. Rubino, C.; Bonet Aracil, M.; Payá, J.G.; Liuzzi, S.; Stefanizzi, P.; Zamorano Cantó, M.; Martellotta, F. Composite eco-friendly sound absorbing materials made of recycled textile waste and biopolymers. *Materials* **2019**, *12*, 4020. [CrossRef] [PubMed]
33. *ISO 12570: 2000. Hygrothermal Performance of Building Materials and Products—Determination of Moisture Content by Drying at Elevated Temperature*; International Organization for Standards: Geneva, Switzerland, 2000.
34. *ASTM D4892-14: 2019. Standard Test Method for Density of Solid Pitch (Helium Pycnometer Method)*; ASTM Internationa: West Conshohocken, PA, USA, 2019.
35. Bouguerra, A.; Aït-Mokhtar, A.; Amiri, O.; Diop, M.B. Measurement of thermal conductivity, thermal diffusivity and heat capacity of highly porous building materials using transient plane source technique. *Int. Commun. Heat Mass* **2001**, *28*, 1065–1078. [CrossRef]
36. *ISO 12572: 2016. Hygrothermal Performance of Building Materials and Products—Determination of Water Vapour Transmission Properties—Cup Method*; International Organization for Standards: Geneva, Switzerland, 2016.
37. *ISO 10534-2: 1998. Acoustics—Determination of Sound Absorption Coefficient and Impedance in Impedance Tubes—Part 2: Transfer-function method*; International Organization for Standards: Geneva, Switzerland, 1998.
38. Delany, M.E.; Bazley, E.N. Acoustical properties of fibrous absorbent materials. *Appl. Acoust.* **1970**, *3*, 105–116. [CrossRef]
39. Miki, Y. Acoustical properties of porous materials—Generalizations of empirical models. *J. Acoust. Soc. Jpn.* **1990**, *11*, 19–24. [CrossRef]
40. Attenborough, K. Acoustical characteristics of porous materials. *Phys. Rep.* **1982**, *82*, 179–227. [CrossRef]
41. Attenborough, K. Acoustical characteristics of rigid fibrous absorbents and granular materials. *J. Acoust. Soc. Am.* **1983**, *73*, 785–799. [CrossRef]
42. Johnson, D.L.; Koplik, J.; Dashen, R. Theory of dynamic permeability and tortuosity in fluid-saturated porous media. *J. Fluid Mech.* **1987**, *176*, 379–402. [CrossRef]
43. Allard, J.F.; Champoux, Y. New empirical equation for sound propagation in rigid frame fibrous material. *J. Acoust. Soc. Am.* **1992**, *91*, 3346–3353. [CrossRef]
44. Lafarge, D.; Lemarinier, P.; Allard, J.F.; Tarnow, V. Dynamic compressibility of air in porous structures at audible frequencies. *J. Acoust. Soc. Am.* **1997**, *102*, 1995–2006. [CrossRef]
45. Pride, S.R.; Gangi, A.F.; Morgan, F.D. Deriving the equations of motion for porous isotropic media. *J. Acoust. Soc. Am.* **1992**, *92*, 3278–3290. [CrossRef]
46. Allard, J.F.; Atalla, N. *Propagation of Sound in Porous Media: Modelling Sound Absorbing Materials*, 2nd ed.; John Wiley & Sons, Ltd.: Chichester, West Sussex, UK, 2009.
47. Ingard, U.K.; Dear, T.A. Measurement of acoustic flow resistance. *J. Sound Vib.* **1985**, *103*, 567–572. [CrossRef]
48. Atalla, N.; Panneton, R. Inverse acoustical characterization of open cell porous media using impedance tube measurements. *Can. Acoust.* **2005**, *33*, 11–24.
49. *JCGM 100:2008. Evaluation of Measurement Data—Guide to the Expression of Uncertainty in Measurement*; International Bureau of Weights and Measures (BIPM): Saint-Cloud, France, 2008. Available online: https://www.bipm.org/en/publications/guides/gum.html (accessed on 20 October 2020).

50. Rode, C.; Peuhkuri, R.; Hansen, K.K.; Time, B.; Svennberg, K.; Arfvidsson, J.; Ojanen, T. NORDTEST project on moisture buffer value of materials. In Proceedings of the AIVC 26th conference: Ventilation in Relation to the Energy Performance of Buildings. Air Infiltration and Ventilation, Brussels, Belgium, 21–23 September 2005; INIVE EEIG: Brussels, Belgium, 2005; pp. 47–52.
51. Muthurajab, R.; Lacoste, C.; Lacroix, P.; Bergeret, A. Sustainable thermal insulation biocomposites from rice husk, wheat husk, wood fibers and textile waste fibers: Elaboration and performances evaluation. *Ind. Crops Prod.* **2019**, *135*, 238–245. [CrossRef]
52. Binici, H.; Aksogan, O. Eco-friendly insulation material production with waste olive seeds, ground pvc and wood chips. *J. Build. Eng.* **2016**, *5*, 260–266. [CrossRef]
53. Pinto, J.; Cruz, D.; Paiva, A.; Pereira, S.; Tavares, P.; Fernandes, L.; Varum, H. Characterization of corncob as a possible raw building material. *Constr. Build. Mater.* **2012**, *34*, 28–33. [CrossRef]
54. Agoudjila, B.; Benchabane, A.; Boudennec, A.; Ibos, L.; Fois, M. Renewable materials to reduce building heat loss: Characterization of date palm wood. *Energy Build.* **2011**, *43*, 491–497. [CrossRef]
55. Nguyen, D.M.; Grillet, A.C.; Diep, T.M.H.; Bui, Q.B.; Woloszyn, M. Characterization of hygrothermal insulating biomaterials modified by inorganic adsorbents. *Heat Mass Transf.* **2020**, *56*, 2473–2485. [CrossRef]
56. Umnova, O.; Attenborough, K.; Shin, H.-C.; Cummings, A. Deduction of tortuosity and porosity from acoustic reflection and transmission measurements on thick samples of rigid-porous materials. *Appl. Acoust.* **2005**, *66*, 607–624. [CrossRef]
57. Glé, P.; Gourdon, E.; Arnaud, L. Acoustical properties of materials made of vegetable particles with several scales of porosity. *Appl. Acoust.* **2011**, *72*, 249–259. [CrossRef]
58. Sarradj, E.; Lerch, T.; Hubelt, J. Input parameters for the prediction of acoustical properties of open porous asphalt. *Acta Acust. United Ac.* **2006**, *92*, 86–96.
59. Ghanbarian, B.; Hunt, A.G.; Ewing, R.P.; Sahimi, M. Tortuosity in porous media: A critical review. *Soil Sci. Soc. Am. J.* **2013**, *7*, 1461–1477. [CrossRef]
60. Turo, D.; Umnova, O. Time Domain Modelling of Sound Propagation in Porous Media and the Role of Shape Factors. *Acta Acust. United Ac.* **2010**, *96*, 225–238. [CrossRef]

materials MDPI

Article

The Influence of Environmentally Friendly Flame Retardants on the Thermal Stability of Phase Change Polyurethane Foams

Dong Liu and Anjie Hu *

School of Civil Engineering and Architecture, Southwest University of Science and Technology, Mianyang 621010, China; dtld123@126.com
* Correspondence: anjie@swust.edu.cn; Tel.: +86-15281112512

Received: 9 December 2019; Accepted: 17 January 2020; Published: 22 January 2020

Abstract: To improve thermal insulation, microencapsulated phase change materials (micro-PCMs), expandable graphite (EG), and ammonium polyphosphate (APP) were introduced into polyurethane foam (PUF) to enhance the thermal stability and improve the thermal insulation behavior. The morphology of the PUF and micro-PCM was studied using a scanning electronic microscope (SEM), while the thermophysical properties of the PUF were investigated using a hot disk thermal constants analyzer and differential scanning calorimetry (DSC). The thermal stability of the PUF was investigated by thermogravimetric analysis (TGA), and the gas products volatilized from the PUF were analyzed by thermogravimetric analysis coupled with Fourier transform infrared spectrometry (TGA-FTIR). The results revealed that the thermal conductivities of the PUF were reduced because micro-PCM is effective in absorbing energy, showing that the PUF functions not only as a thermal insulation material but also as a heat sink for energy absorption. Moreover, the EG and APP were found to be effective in improving the thermal stabilities of the PUF, and the optimized formulation among EG, APP, and micro-PCMs in the PUF showed a significant synergistic effect.

Keywords: polyurethane foam; thermal property; phase change material; flame retardant

1. Introduction

Widely used for sound absorption [1–3], furniture [3,4], and insulating materials [5], polyurethane foam (PUF) plays an important role in our daily life. Compared with other insulating materials, PUF has the advantage of low thermal conductivity, high mechanical and chemical stability at both high and low temperatures, and the ability to form sandwich structures with various facer materials [6,7].

The properties of PUF depend on the density, structure, and geometry of the foam, and a large number of fillers are required to get the desired properties [8–10]. To improve the thermal characteristics of PUF, adding phase change material (PCM) into PUF has been studied since the 1990s [11–13], since PCM can absorb or release significant latent heat while the temperature of the material stays almost constant in the process of phase change [11,14]. Among the investigated PCMs, the paraffin shows desirable properties, such as high heat of fusion, little or no supercooling, self-nucleating behavior, and thermal and chemical stability. However, liquid paraffin has a relatively low viscosity, and it has to be kept in a closed tank or container to prevent leaching. Microencapsulation of paraffin provides a perfect way to solve the leaching problem. Microcapsules containing the polar phase change material (PCM) n-dodecanol were synthesized by in situ polymerization, using melamine-formaldehyde resin as a shell and styrene-maleic anhydride copolymer (SMA) as an emulsifier. Results show that anionic SMA emulsifier is suitable for the encapsulation of n-dodecanol [15]: when the mass ratio of emulsifier to n-dodecanol is 4.8%, the phase change latent heat and encapsulation efficiency reaches the maximum values of 187.5 J/g and 93.1%, respectively. A series of PU microcapsules containing the PCM of n-octadecane was successfully synthesized, using an interfacial polymerization in SMA dispersion with diethylene triamine (DETA) as a chain extender reacting with toluene-2,4-diisocyanate (TDI), and the average diameter of the microencapsulated phase change materials (micro-PCMs) was in the range of 5–10 μm. Earlier research showed that PU-shells regularly fabricated with the influence of SMA varied with a weight ratio from 1.0% to 4.0% in diameter, encapsulating ratio, release properties, and thermal stability [16].

Another drawback of PUF and paraffin is that, like other organic materials, their flammability is high and the flame-spreading rate is fast, which limits their use due to the high possibility of a fire hazard. Hence, a flame retardant should be added to improve the thermal stability of PUF and paraffin. There are generally two kinds of flame retardants. The first are halogen-containing flame retardants, which are most effective at improving thermal stability. However, they always release a large amount of noxious material during their decomposition, which has brought ecological and physical problems. The second kind are halogen-free flame retardants, such as EG and decabrominated dipheny ethane (DBDPE). The effect of the retardants is influenced by many factors. In the literature [17], the EG/PU composites prepared with their method showed a V-0 flame retardance level, whereas EG/PU composite prepared by conventional blending only showed a V-2 flame retardance level. Results in the literature [9] showed that the flame-retardant efficiency got better with the increase in the density of PU foam at the fixed EG weight percent, or with an increase in the EG weight percent at the fixed foam density. According to the literature [8], when the flame-retardant loadings were 20 wt%, the limiting oxygen index (LOI) value of DBDPE-filled rigid polyurethane foam (RPUF) increased to 33 vol%, while the EG-filled RPUF reached 41%. However, when they were simultaneously added into RPUF, there was not any flame-retardant synergistic effect. In the literature [18], they tested polystyrene composites with 21.4 wt% ammonium polyphosphate, and found that the peak of the heat release rate was reduced by 77.5% as compared to those of virgin polystyrene. In the literature [19], it was found that the peak heat release rate of PEG could be decreased by 19.2% by introducing graphene aerogels (GA) with 1.60 wt% and phosphorylated polyvinyl alcohol (PPVA) with 15.0 wt%. In the literature [20], it was found that, with 0.25 phr of functionalized graphene oxide (fGO) and 10 phr of EG/dimethyl methyl phosphonate (DMMP), the LOI value of RPUF/fGO composites reached 28.1%, and its UL-94 test reached V-0 rating, showing that the fGO could significantly enhance the flame retardant properties of RPUF composites. Carbon nanotubes and zinc aluminum-layered double hydroxide (CNT/ZnAl-LDH) could improve the thermal degradation stability as well as the fire safety of PU foams [21]. These researches indicate that halogen-free flame retardants can effectively improve the stability of the organic materials, and their effects are mostly influenced by the content of the retardants and the synergistic effect of different retardants. Among these retardants, the EG and APP are widely applied due to their good flame-retardant efficiency and economic efficiency. Thus, we applied them in the PUF, and optimized the content of the retardants for the best effect.

In this work, we focused on the effects of EG, APP, and micro-PCMs on the thermal characteristics of PUF. The EG and the APP were added to increase the thermal stability of PUF synergistically, and micro-PCMs were added to improve the temperature-controlling capability of PUF. Besides thermal stability, the EG can also improve the effectiveness of the energy exchange of PCM because of its high thermal conductivity.

2. Experimental Section

2.1. Materials

Paraffin (melting temperature T_m = 58.15 °C, latent heat 163.31 J/g) was used as the phase change material. The melamine (MA), formaldehyde, and dodecyl sodium sulfate were supplied by Sinopharm Chemical Ltd. (Shanghai, China). The polymeric MDI (methane diphenyl diisocyanate, NCO% = 30.2~32.5, viscosity at 25 °C equals to 150~250 mPa·s) was purchased from Yantai Wanhua

Polyurethanes Co., Ltd. (Yantai, China). The polyether polyol (YD 4450, n°OH = 350 ± 10 mgKOH/g, viscosity at 25 °C equals to 220 ± 50 MPa·s) was supplied by Hebei Yadong Chemical Group Co., Ltd. (Shijiazhuang, China). The ammonium polyphosphate (APP, average particle size 92% < 10 μm) and expandable graphite (EG, average particle size: 320 mesh) were provided by Hefei Keyan Chemical Material Technology Development Co., Ltd (Hefei, China). Other additives are given as follows: dibutyltin dilaurate (DD), triethylenediamine, triethanolamine, silicone oil, and water.

2.2. Preparation of Micro-PCMs

Micro-PCMs were fabricated by in situ polymerization, using a melamine–formaldehyde copolymer as the shell and paraffin as the core material. The detailed fabrication process included the steps of prepolymer solution synthesis, PCM emulsion preparation, and mixing of the prepolymer solution and the PCM emulsion for an in situ polymerization. The resultant microcapsules were filtered and washed with distilled water at 70 °C until a pH of 7 was reached. The wet powders were then dried in a vacuum oven at 80 °C for 12 h to remove the water. The final mass fraction of the paraffin was approximately 70%, and the total latent heat of the micro-PCM was 118.9 J/g.

2.3. Fabrication of PUF

The PUF formulations are given in Table 1. Making the PUF, materials of catalysts, surfactants, blowing agents, EG, APP, and micro-PCMs, etc. were added to a YD 4450 blend in a proper mold and stirred intensively for 15 s. After that, the calculated quantity of MDI was added and stirred at 2500 rpm for 10 s at a room temperature of 20 °C. Then the foam was placed in an oven at 60 °C for 24 h.

Table 1. Formulations and thermophysical properties of PUFs.

	MDI (g)	YD-4450 (g)	EG (g)	APP (g)	Micro-PCM (g)	Energy Storage Capacity (J/g)	Energy Release Capacity (J/g)
PU1	50	50	0	0	0	0	0
PU2	50	50	10	0	0	0	0
PU3	50	50	15	0	0	0	0
PU4	50	50	8	2	0	0	0
PU5	50	50	13	2	0	0	0
PU6	50	50	10	0	10	7.73	5.87
PU7	50	50	15	0	10	6.58	5.55
PU8	50	50	8	2	10	8.21	7.12
PU9	50	50	13	2	10	6.66	6.02

2.4. Characterization of Flame Retardant PUF

The morphology and microstructure of the PUF were observed using an electronic scanning microscope (Hitachi X650, Hitachi Limited, Tokyo, Japan). The latent heat and the phase change temperature of the PUF were studied using differential scanning calorimetry (DSC), which was carried out in an argon atmosphere with the DT-50 (Shimadzu, Quioto, Japan) thermal analyzer, with a heating rate of 10 °C/min and a flow rate of 30 mL/min. The thermal conductivities of the PUF were tested by the hot disk thermal constants analyzer. A constant voltage was applied, and a thermocouple recorded the temperature rise by circulating a current along the platinum wire. The precision of the thermal conductivity and the thermal diffusivity were ± 2.0% and ± 5.0%, respectively. The thermal stability properties of the PUFs were tested by a TGA Q5000 IR thermogravimetric analyzer (TA Instruments, New Castle, PA, USA). In the test, the samples were heated from room temperature to 600 °C with a linear heating rate of 20 °C/min in a nitrogen atmosphere. The thermogravimetric analysis-Fourier transform infrared spectrometry (TGA-FTIR) of the sample was performed using the DT-50 instrument interfaced to the Varian 2000 FTIR spectrometer. About 3–5 mg of the product was put in an alumina crucible and heated from 30 to 600 °C. The heating rate was set as 20 °C/min in a nitrogen atmosphere.

3. Results and Discussion

3.1. SEM Photos of Micro-PCMs and Pure PUF Structure

Figure 1a shows the particle profiles of the micro-PCMs. The micro-PCMs have a spherical shape, and the microcapsules range in size from 0.3 to 0.8 μm. The entire honeycomb structure of the pure PUF is shown in Figure 1b, and the honeycomb structure can contain the additives, such as EG, APP, and micro-PCMs. Observing the PUF (Figure 1c), it can be found that the admixture inserted into the pure PUF, showing that the PUF could be used as a supporting material for the admixture.

Figure 1. SEM photographs of the micro-PCM (**a**) and PUF (**b**,**c**).

3.2. The Thermal Conductivities of the PUFs

As a widely used thermal insulation material, the heat transfer efficiency of the PUF is the most important property in the application of the material [22,23]. The heat transfer efficiency of PUF is characterized by thermal conductivity [24]. The tested thermal conductivities of PUFs with different formulation are listed in Figure 2, which show that both the additions of EG and APP increase the thermal conductivities of PUFs. However, the increases in the conductivities are relatively small compared to the conductivity of the pure PUF. According to the heat transfer theory, the thermal conductivities of PUFs are determined by the thermal conductivity of the foams and gases within the cell, the radiative heat transfer among these foams, and the convection heat transfer within the cells. Filler particles, including EG and APP, not only fit into the structure of the foam but also lie within the cells, which increases the thermal conductivity of foams and cell gases. However, the filler particles also decrease the radiative and convection heat transfer within the cells, which reduces the effects of the heat transfer enhancement. This conclusion also can explain the result that the thermal conductivities change indistinctively when increasing the content of EG in the PUF. Double side effects also exist in adding the micro-PCMs. When the micro-PCMs are added to the PUF, the thermal conductivities increase, due to their capability for absorbing energy. However, this effect does not reduce the thermal insulation of the PUF, because the energy is restricted into the phase change material.

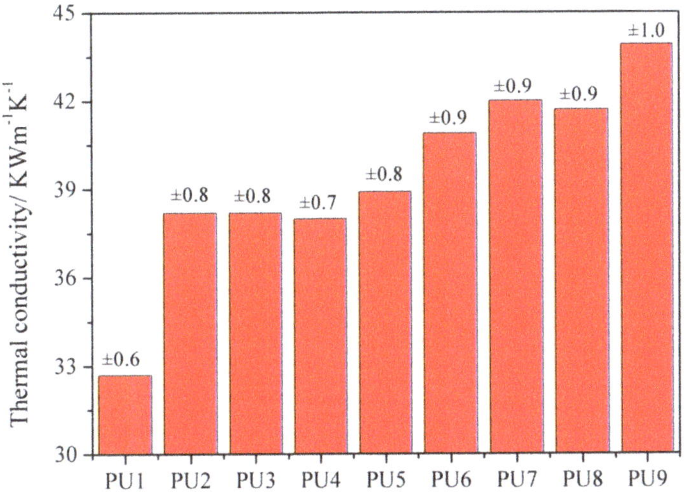

Figure 2. The thermal conductivities of PUFs.

3.3. The Latent Heats of the PUFs

The latent heats of the PUFs are tested by the DSC. The DSC curves of the micro-PCM and PU6 to PU9 are exhibited in Figures 3 and 4, respectively. It can be seen that there are two peaks on the DSC curves. The sharp peak represents the solid–liquid phase change of the paraffin, and the minor peak is associated with the liquid–solid phase transition of paraffin. The values of the latent heat obtained with DSC curves are presented in Table 1. As shown in the table, the total latent heats of the PUFs are a little lower than the theoretical values, which were calculated by multiplying the latent heat of the dispersed micro-PCMs with their weight percentage. DSC is also used to investigate the changes of latent heat values under conditions of thermal cycling with the temperature from 80 to 25 °C. From Figure 4 and Table 1, we can find that the energy was released when the temperature decreased, which is useful in order to keep a constant temperature for environment and energy saving. These results demonstrate that the PUFs are not only thermal insulation materials but also storage materials for thermal energy.

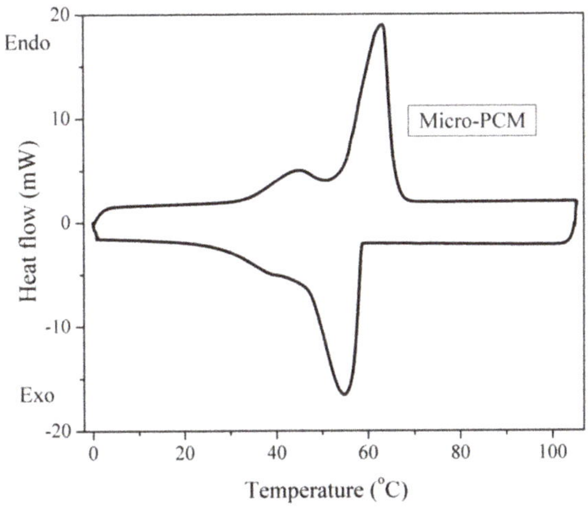

Figure 3. The DSC curves during thermal cycling of micro-PCM.

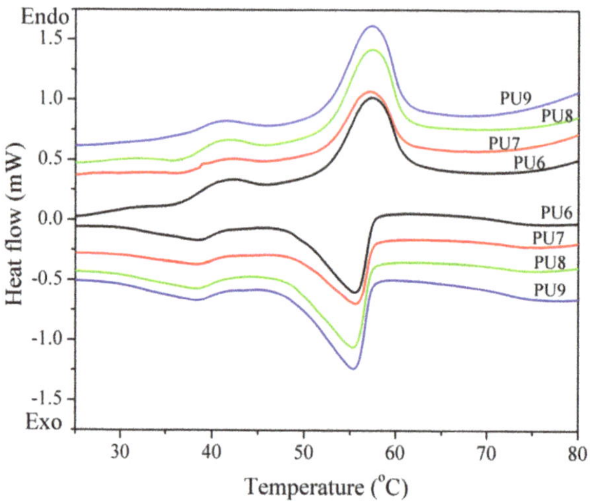

Figure 4. The DSC curves during thermal cycling of PUFs.

3.4. The Thermal Stabilities of PUFs

The thermal stability of the PUF as a thermal insulation material is important for practical applications. Since PUF and micro-PCM both have low thermal stability, they are ignited easily. Improving the thermal stability of PUF is imperative. Figures 5 and 6 show the thermograms of micro-PCMs and PUFs, respectively. The thermal degradation of pure PUF occurs in a two-step process. The first step of degradation is due to the formation of polyol and isocyanate groups. The isocyanates from this deployed condensation reaction are very reactive to dimerize, and they form carbodiimide with the evolution of carbon dioxide. The second step of degradation was due to the further reaction between carbodiimide and polyol to impart the substituted urea [25,26]. From Figure 6a, it can be found that there are no significant changes for the thermal degradation of the PU2 and PU3 when the EG is introduced into the PU1. This is probably due to no reaction/interaction between EG and PUF during the thermal degradation process. However, the rates of thermal degradation for the PU2 and PU3 are accelerated at about 300 to 400 °C in comparison with that of PU1, which may be due to the degradation of intercalated compounds in EG, such as H_2SO_4 decomposing to H_2O and SO_2 gases and the oxidation of EG, and the distance between the basal planes of EG increasing [27,28]. The structure of EG could prevent the heat and oxygen transfer and improve the thermal stability of the PUF at high temperatures because the EG could cover the surface of the matrix. At higher loadings of EG in PUF, the content of the char layer increases in the temperature range from about 650 to 750 °C, and this char layer could act as a thermal barrier for the PUF and prevent further decomposition. When the mass fraction of APP replaces the EG in equal amounts of PU2 and PU3, the thermal stabilities of the PU4 and PU5 are both increased. This is because the APP can form polyphosphoric acid and carbonaceous residue with PUF at high temperatures, and help in inhibiting the thermal degradation of the PUF. When the micro-PCMs are introduced into the PU2 and PU3, the thermal stabilities of the PU6 and PU7 are both lower than that of PU2 and PU3 (Figure 6b). This is because the thermal stability of the micro-PCMs (listed in Figure 5) is lower than that of PU2 and PU3. However, if part of the EG of PU6 and PU7 is replaced with equal amounts of the mass fraction of APP, the thermal stability of PU8 becomes higher than that of PU6 (Figure 6c), while the PU9 degrades faster than PU7 for temperatures lower than 550 °C and slower for higher temperatures, showing that the ingredients influence the thermal stability improvement due to the synergistic effect of APP and EG. These results show that the APP, melamine and polyol can be effectively taken as an integrated intumescent flame retardant, and the intumescent flame retardant can be thermally decomposed at a higher temperature to form three-dimensional network structures. Besides, the EG acted in great synergistic effect with the APP with a certain formulation in the PUF/micro-PCM system [29].

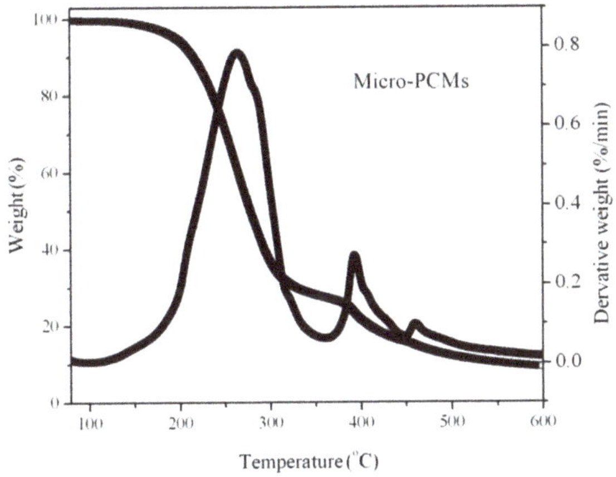

Figure 5. The thermal stability of micro-PCMs.

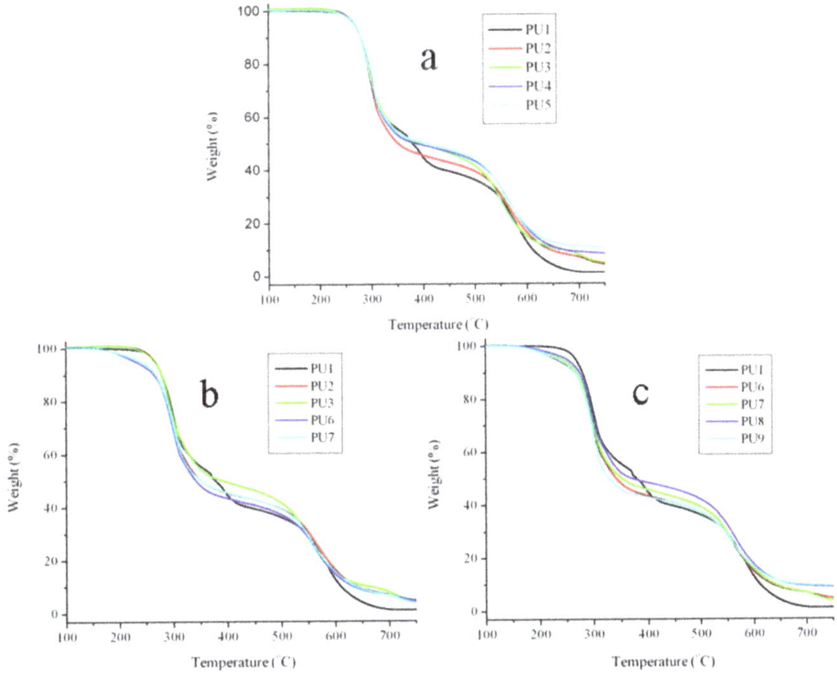

Figure 6. The thermal stabilities of PUFs (**a**) the effects of EG/APP on the thermal stability of PUF, (**b**) the effects of micro-PCMs on the thermal stability of PUF/EG composite, (**c**) the effects of EG/APP on the thermal stability of PUF/micro-PCMs composite.

3.5. The Volatilized Products of the PUFs

The TGA-FTIR study is to identify the volatilized products, and the results can provide a better understanding of the thermal degradation mechanisms [30]. The whole range of spectra of the samples is shown in Figure 7. It can be seen that the main decomposition product is carbon dioxide, which is in the range of 2410–2220 cm^{-1}. In order to investigate the influence of EG, APP, and micro-PCMs on the thermal degradation of the PUF, the evolution of the intensity of the characteristic peak carbon dioxide is plotted in Figure 8, in which the characteristic peak is at 2356 cm^{-1}. It can also be found that the curves of the carbon dioxide show two peaks at about 16 and 32 min, respectively. These peaks are attributed to the depolymerization of PUF to form monomer precursors such as polyol and isocyanates, and the dimerization from isocyanate to carbodiimide and carbon dioxide. The ratio between the second peak value and the first peak value are calculated, i.e., the ratio of PU1 is 0.71, while the ratios of the PU2, PU4, PU6, and PU8 are 1.09, 2.10, 2.84, and 2.23, respectively, showing that EG can highly reduce the carbon dioxide. Comparing the results of PU2 and PU6 with PU4 and PU8, we can also find the reduction of carbon dioxide when replacing EG with APP in PUF. From these results, it can be concluded that EG increased the thermal stability of the PUF, and by replacing part of EG with APP, the efficiency in improving the thermal stability of PU can be further improved.

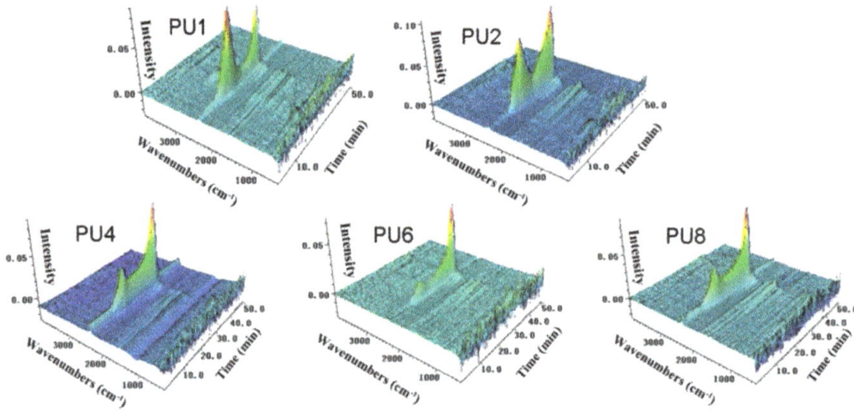

Figure 7. The 3D surface graphs of the FTIR spectra of the evolved gases produced by PUFs.

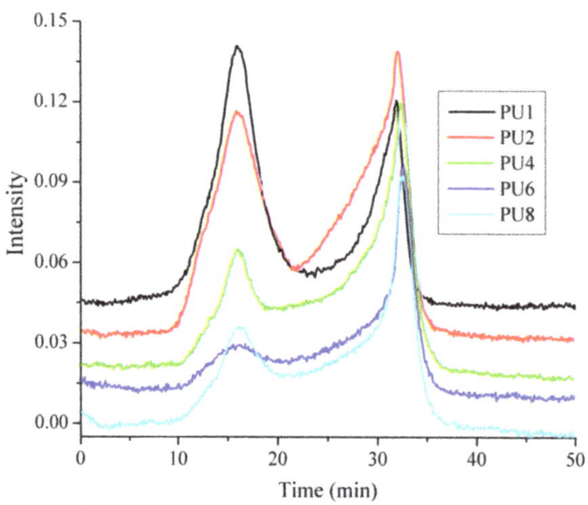

Figure 8. Intensity of characteristic peak versus time for the volatilized carbon dioxide of PUFs.

4. Conclusions

The preparation and characterization of PUF containing EG, APP, and micro-PCM are reported. EG and APP were used to improve the thermal stability of PUF, and micro-PCMs were used as the phase change material for thermal energy storage. The SEM images showed that PUF could be an excellent supporting material for EG, APP, and micro-PCM. The thermal conductivity results indicated that the thermal conductivities of the PUF changed owing to the added EG and APP, but the values do not increase continually with the increase of EG and APP contents in the PUF. A possible reason is that the EG and APP could increase the thermal conductivities of the foams and cell gases while decreasing the radiative heat transfer within the foams. The thermal conductivities increased when micro-PCMs were introduced into the PUF because micro-PCMs could absorb the thermal energy; hence, the temperature could be maintained more consistently by the PUF. The DSC results showed that the total latent heats of the PUFs were a little lower than the theoretically calculated values. The TGA and TGA-FTIR results showed that APP and EG have synergistic effects for improving the thermal stabilities of the PUF. When the micro-PCMs are added into PUF, an integrated intumescent flame-retardant system is formed at high temperatures, and optimal formulation among EG, APP, and micro-PCM could provide the optimal synergistic effect in improving the thermal stability of the PUF. All in all, the temperature regulating and controlling effect could be ameliorated owing to micro-PCM, the thermal conductivity of PUF improved with the appropriate amount of EG, and EG and APP showed a great synergistic effect to improve the thermal stability of the PUF/micro-PCM composite.

Author Contributions: Conceptualization, D.L. and A.H.; methodology, A.H.; software, D.L.; validation, A.H.; formal analysis, D.L.; investigation, D.L.; resources, D.L. and A.H.; data curation, D.L.; writing—original draft preparation, D.L.; writing—review and editing, A.H.; visualization, D.L.; supervision, A.H.; project administration, A.H.; funding acquisition, A.H. and D.L. All authors have read and agreed to the published version of the manuscript.

Funding: This research was funded by National Natural Science Foundation of China, grant number 51503173 and 51306156; Sichuan Provincial Department of Education, grant number 17ZA0405; Doctoral Program of Southwest University of Science and Technology, grant number 11zx7156; The Open Fund of Jinhe, grant number 11zh9002. The APC was funded by Sichuan Provincial Department of Education, grant number 17ZA0405.

Conflicts of Interest: The authors declare no conflict of interest. The funders had no role in the design of the study; in the collection, analyses, or interpretation of data; in the writing of the manuscript, or in the decision to publish the results.

References

1. Li, T.T.; Zhang, X.; Wang, H.Y.; Dai, W.N.; Huang, S.Y.; Shiu, B.C.; Lou, C.W.; Lin, J.H. Sound absorption and compressive property of PU foam-filled composite sandwiches: Effects of needle-punched fabric structure, porous structure, and fabric-foam interface. *Polym. Adv. Technol.* **2019**. [CrossRef]
2. Tiuc, A.E.; Nemes, O.; Vermesan, H.; Toma, A.C. New sound absorbent composite materials based on sawdust and polyurethane foam. *Compos. Part B Eng.* **2019**, *165*, 120–130. [CrossRef]
3. Park, M.; Park, H.K.; Shin, H.K.; Kang, D.; Pant, B.; Kim, H.; Song, J.K.; Kim, H.Y. Sound absorption and insulation properties of a polyurethane foam Mixed with electrospun nylon-6 and polyurethane nanofibre Mats. *J. Nanosci. Nanotechnol.* **2019**, *19*, 3558–3563. [CrossRef]
4. Cooper, E.M.; Kroeger, G.; Davis, K.; Clark, C.R.; Ferguson, P.L.; Stapleton, H.M. Results from screening polyurethane foam based consumer products for flame retardant chemicals: Assessing impacts on the change in the furniture flammability standards. *Environ. Sci. Technol.* **2016**, *50*, 10653–10660. [CrossRef] [PubMed]
5. Liu, L.; Wang, Z.Z.; Zhu, M.H. Flame retardant, mechanical and thermal insulating properties of rigid polyurethane foam modified by nano zirconium amino-tris-(methylenephosphonate) and expandable graphite. *Polym. Degrad. Stabil.* **2019**, *170*, 108997. [CrossRef]
6. Harikrishnan, G.; Singh, S.N.; Kiesel, E.; Macosko, C.W. Nanodispersions of carbon nanofiber for polyurethane foaming. *Polym.* **2010**, *51*, 3349–3353. [CrossRef]
7. Dementyev, A.G.; Dementyev, M.A.; Zinger, P.A.; Metlyakova, I.R. Effect of the cellular structure on thermal conductivity of rigid closed-cell foam polymers during long-term aging. *Mech. Compos. Mater.* **1999**, *35*, 129–138. [CrossRef]
8. Ye, L.; Meng, X.Y.; Liu, X.M.; Tang, J.H.; Li, Z.M. Structure and magnetic properties of regenerated cellulose/Fe3O4 nanocomposite films. *J. Appl. Polym. Sci.* **2009**, *111*, 2372–2380. [CrossRef]

9. Bian, X.C.; Tang, J.H.; Li, Z.M.; Lu, Z.Y.; Lu, A. Dependence of flame-retardant properties on density of expandable graphite filled rigid polyurethane foam. *J. Appl. Polym. Sci.* **2007**, *104*, 3347–3355. [CrossRef]
10. Thirumal, M.; Khastgir, D.; Singha, N.K.; Manjunath, B.S.; Naik, Y.P. A facile approach to incorporate silver nanoparticles into dextran-based hydrogels for antibacterial and catalytical application. *J. Macromol. Sci. A.* **2009**, *46*, 704–712. [CrossRef]
11. Mulligan, J.C.; Colvin, D.P.; Bryant, Y.G. Microencapsulated phase-change material suspensions for heat transfer in spacecraft thermal systems. *J. Spacecr. Rockets.* **1996**, *33*, 278–284.
12. Sarier, N.; Onder, E. Thermal characteristics of polyurethane foams incorporated with phase change materials. *Thermochim. Acta.* **2007**, *454*, 90–98. [CrossRef]
13. You, M.; Zhang, X.X.; Li, W.; Wang, X.C. Effects of MicroPCMs on the fabrication of MicroPCMs/polyurethane composite foams. *Thermochim. Acta.* **2008**, *472*, 20–24. [CrossRef]
14. Pasupathy, A.; Athanasius, L.; Velraj, R.; Seeniraj, R.V. Experimental investigation and numerical simulation analysis on the thermal performance of a building roof incorporating phase change material (PCM) for thermal management. *Appl. Therm. Eng.* **2008**, *28*, 556–565. [CrossRef]
15. Yu, F.; Chen, Z.H.; Zeng, X.R. Preparation characterization and thermal properties of microPCMs containing n-dodecanol by using different types of styrene-maleic anhydride as emulsifier. *Colloid Polym. Sci.* **2009**, *287*, 549–560. [CrossRef]
16. Su, J.F.; Wang, L.X.; Ren, L. Synthesis of polyurethane microPCMs containing n-octadecane by interfacial polycondensation: Influence of styrene-maleic anhydride as a surfactant. *Colloid Surface A.* **2007**, *299*, 268–275. [CrossRef]
17. Zhang, C.; Shi, M.N.; Zhang, Y.C.; Yang, W.M.; Jiao, Z.W.; Yang, L.P. EG/TPU Composites with enhanced flame retardancy and mechanical properties prepared by microlayer coextrusion technology. *RSC Adv.* **2019**, *9*, 23944–23956. [CrossRef]
18. Shi, Y.Q.; Gui, Z.; Yuan, B.H.; Hu, Y.; Zheng, Y.Y.J. Flammability of polystyrene/aluminim phosphinate composites containing modified ammonium polyphosphate. *Therm. Anal. Calorim.* **2018**, *131*, 1067–1077. [CrossRef]
19. Shen, J.; Zhang, P.; Song, L.X.; Li, J.P.; Ji, B.Q.; Li, J.J.; Chen, L. Polyethylene glycol supported by phosphorylated polyvinyl alcohol/graphene aerogel as a high thermal stability phase change material. *Compos. Part B Eng.* **2019**, *179*, 107545. [CrossRef]
20. Gao, M.; Li, J.F.; Zhou, X. A flame retardant rigid polyurethane foam system including functionalized graphene oxide. *Polym. Compos.* **2019**, *40*, 1274–1282. [CrossRef]
21. Elbasuney, S. Surface engineering of layered double hydroxide (LDH) nanoparticles for polymer flame retardancy. *Powder Technol.* **2015**, *277*, 63–73. [CrossRef]
22. Santiago-Calvo, M.; Tirado-Mediavilla, J.; Ruiz-Herrero, J.L.; Villafane, F.; Rodriguez-Perez, M.A. Long-term thermal conductivity of cyclopentane¿water blown rigid polyurethane foams reinforced with different types of fillers. *Polym. Int.* **2019**, *68*, 1826–1835. [CrossRef]
23. Wang, F.; Zhang, P.; Mou, Y.R.; Kang, M.; Liu, M.; Song, L.X.; Lu, A.; Rong, J.Z. Synthesis of the polyethylene glycol solid-solid phase change materials with a functionalized graphene oxide for thermal energy storage. *Polym. Test.* **2017**, *63*, 494–504. [CrossRef]
24. Yan, Y.J.; Xia, H.; Qiu, Y.P.; Xu, Z.Z.; Ni, Q.Q. Fabrication of gradient vapor grown carbon fiber based polyurethane foam for shape memory driven microwave shielding. *RSC Adv.* **2019**, *9*, 9401–9409. [CrossRef]
25. Chattopadhyay, D.K.; Webster, D.C. Thermal analysis of polyurethane dispersions based on different polyols. *Prog. Polym. Sci.* **2009**, *34*, 1068–1133. [CrossRef]
26. Tang, Z.; Maroto-Valer, M.M.; Andresen, J.M. Thermal degradation behavior of rigid polyurethane foams prepared with different fire retardant concentrations and blowing agents. *Polym.* **2002**, *43*, 6471–6479. [CrossRef]
27. Duquesne, S.; Le Bras, M.; Bourbigot, S.; Delobel, R.; Vezin, H.; Camino, G.; Eling, B.; Lindsay, C.; Roels, T. Expandable graphite: A fire retardant additive for polyurethane coatings. *Fire Mater.* **2003**, *27*, 103–117. [CrossRef]
28. Duquesne, S.; Delobel, R.; Le Bras, M.; Camino, G. A comparative study of the mechanism of action of ammonium polyphosphate and expandable graphite in polyurethane. *Polym. Degrad. Stabil.* **2002**, *77*, 333–344. [CrossRef]

29. Zheng, Z.H.; Liu, Y.; Zhang, L.; Wang, H.Y. Synergistic effect of expandable graphite and intumescent flame retardants on the flame retardancy and thermal stability of polypropylene. *J. Mater. Sci.* **2016**, *51*, 5857–5871. [CrossRef]
30. Karimpour-Motlagh, N.; Khonakdar, H.A.; Jafari, S.H.; Panahi-Sarmad, M.; Javadi, A.; Shojaei, S.; Goodarzi, V. An experimental and theoretical mechanistic analysis of thermal degradation of polypropylene/polylactic acid/clay nanocomposites. *Polym. Adv. Technol.* **2019**, *30*, 2695–2706. [CrossRef]

materials

MDPI

Review

Acoustic Properties of Innovative Concretes: A Review

Roman Fediuk [1], Mugahed Amran [2,3,*], Nikolai Vatin [4], Yuriy Vasilev [5], Valery Lesovik [6,7] and Togay Ozbakkaloglu [8]

1 Polytechnic Institute, Far Eastern Federal University, 690922 Vladivostok, Russia; roman44@yandex.ru
2 Department of Civil Engineering, College of Engineering, Prince Sattam Bin Abdulaziz University, Alkharj 11942, Saudi Arabia
3 Department of Civil Engineering, Faculty of Engineering and IT, Amran University, Amran 9677, Yemen
4 Peter the Great St. Petersburg Polytechnic University, 195251 St. Petersburg, Russia; vatin@mail.ru
5 Department of Road-Building Materials, Moscow Automobile and Road Construction University, 125319 Moscow, Russia; yu.vasilev@madi.ru
6 Belgorod State Technological University named after V. G. Shoukhov, 308012 Belgorod, Russia; naukavs@mail.ru
7 Research Institute of Building Physics, The Russian Academy of Architecture and Building Sciences, 127238 Moscow, Russia
8 Ingram School of Engineering, Texas State University, San Marcos, TX 78667, USA; togay.oz@txstate.edu
* Correspondence: m.amran@psau.edu.sa

Abstract: Concrete is the most common building material; therefore, when designing structures, it is obligatory to consider all structural parameters and design characteristics such as acoustic properties. In particular, this is to ensure comfortable living conditions for people in residential premises, including acoustic comfort. Different types of concrete behave differently as a sound conductor; especially dense mixtures are superior sound reflectors, and light ones are sound absorbers. It is found that the level of sound reflection in modified concrete is highly dependent on the type of aggregates, size and distribution of pores, and changes in concrete mix design constituents. The sound absorption of acoustic insulation concrete (AIC) can be improved by forming open pores in concrete matrices by either using a porous aggregate or foam agent. To this end, this article reviews the noise and sound transmission in buildings, types of acoustic insulating materials, and the AIC properties. This literature study also provides a critical review on the type of concretes, the acoustic insulation of buildings and their components, the assessment of sound insulation of structures, as well as synopsizes the research development trends to generate comprehensive insights into the potential applications of AIC as applicable material to mitigate noise pollution for increase productivity, health, and well-being.

Keywords: concrete; noise; acoustic properties; sound-absorbing; sound-reflecting

Citation: Fediuk, R.; Amran, M.; Vatin, N.; Vasilev, Y.; Lesovik, V.; Ozbakkaloglu, T. Acoustic Properties of Innovative Concretes: A Review. *Materials* **2021**, *14*, 398. https://doi.org/10.3390/ma14020398

Received: 30 November 2020
Accepted: 13 January 2021
Published: 14 January 2021

1. Introduction

Building regulations and planning authorities have not properly focused on resolving the ongoing issue of noise pollution faced by the residents of urban areas [1]. Typically, concrete is used as an external cladding to inhibit the propagation of sound transmission, which reflects the sound waves away from the structure [2]. Although the sound waves get reflected away, its magnitude does not reduce and becomes an issue in enclosed spaces such as apartment complexes, factories, and narrow thoroughfares, respectively (see Figure 1) [3,4]. As a result, this leads to several problems such as masked warning signals, higher chances of impaired hearing, and increased work-related stresses [5].

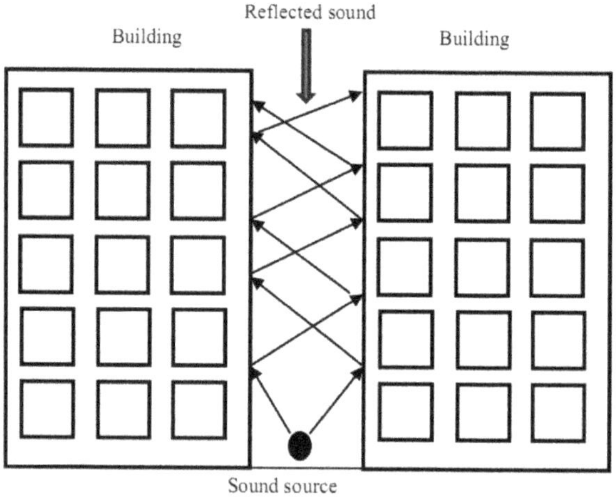

Figure 1. Sound waves reflections in a surrounded narrow thoroughfare [3]. Reprinted with permission from Elsevier [3].

Concrete is a widespread and cheap structural material for the construction of residential buildings of any volume and number of storeys [6]. During the construction and operation of these buildings, it is necessary to ensure comfortable living conditions for people [7]. The most important step along this path is ensuring the acoustic well-being of residents through the optimal design of building envelopes [8–10]. Usually, when a sound wave strikes or hits the surface of a building, the sound energy gets absorbed, transmitted, or reflected away [11]. Hence, there is a relationship between the incident energy and the absorbed, transmitted, or reflected energy [12]. The absorption of sound in a porous material is related to the loss in energy wherein the incident noise energy converts to heat energy and some other energies due to vibration, friction, and air viscosity [13,14]. Sound insulation is the sound transmission loss of the building partition [15,16]. Lightweight porous materials absorb sound well, while heavyweight and hard materials reflect a significant portion of sound energy [3]. Despite the fact that there are some studies on the acoustic behavior of different concrete material, it is necessary to systematize the current state of sound-insulation concrete composites and structures made of them [17,18].

To reduce build time, labor savings, and thermal efficiency, most architects and designers prefer the use of insulated concrete forms. However, the acoustic properties of these insulated concrete forms are often overlooked or under looked. To provide better acoustic properties, the use of acoustic insulation concrete (AIC) is a recent advancement in the field of concrete technology. The acoustic component installed in an AIC reduces the sound in the building structure and keeps the pollution created by noise at bay while improving the acoustical effects of the sound generated within. To achieve a desired acoustic protection, the installation of an AIC wall can provide excellent exterior insulation as well as a soundproofing envelope. Due to a better soundproofing envelope, AIC becomes a suitable choice for hotels, airports, and schools where the requirement of thick traditional walls can be eliminated. Meanwhile, the use of lightweight concretes such as foam or fabric are usually porous and do not allow sound to be reflected. As a result, the sound passes through the porous medium, and the energy gets converted to heat while reducing its magnitude [3]. Hence, the use of lightweight concrete is effective when used in movie halls or in recording studios. This reduces the resonance time in the room [19,20]. Although the use of lightweight concrete is effective internally, it is not suitable in external conditions, and hence, concrete is still the preferred material for other commercial and residential building structures. Based on the past researchers, it was reported that various types of concrete behave differently as a sound conductor; especially dense mixtures are superior sound reflectors, and light ones are sound absorbers. It is found that the level of sound reflection in modified concrete depends on the type of aggregates, size, and distribution of pores and changes in concrete mix design constituents. The sound absorption of AIC is enhanced by forming open pores in concrete matrices by either using a porous aggregate or foam agent. To this end, this article reviews the noise and sound transmission in buildings, types of acoustic insulating materials, and the acoustic properties. This literature study also delivers a comprehensive review on the type of concretes, the acoustic insulation of buildings and their components, the assessment of sound insulation of structures, as well as to synopsize the research development trends to generate comprehensive insights into the potential applications of AIC as applicable material to mitigate noise pollution for increase productivity, health, and well-being.

2. Transmission of Noise and Sound in Buildings

To produce a noise and sound-proof AIC, the walls and roofs of a building could be engineered with modifications on the level at which the sound energy (Figure 2a) [21]:

- Reflects back from the building wall,
- Absorbs within the wall, and
- Transmits through the building wall.

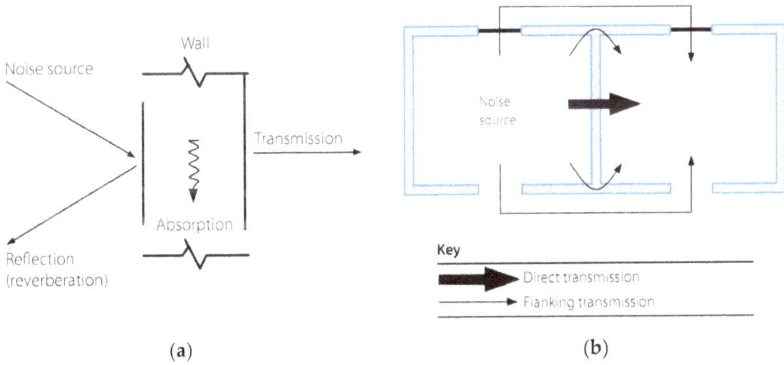

Figure 2. Noise and sound transmission in a building. (**a**) Factors of acoustic insulation performance; (**b**) Flanking and direct transmissions of sound between different zones [21]. Reprinted with permission from DATA STEEL [21].

Structural elements such as glazing and roof lights, which can reduce the performance, should also be considered. The transmission of noise between different zones of a building structure is shown in Figure 2b. Two types of noise transmission occur: direct and flanking transmission, and noise is transmitted when any two areas are separated from each other. The noise travels in two separate routes [22,23]: (i) through the separation structure, which is termed direct transmission, and (ii) around the separation structure through adjoining structural elements, which is termed flanking transmission [21,24]. Meanwhile, the insulation of sound for both types of transmission is controlled by its mass, isolation, and sealing characteristics [22,25]. The direct transmission of noise can be determined by laboratory tests, since it directly depends on the properties of the separation wall and floor [26]. Meanwhile, the details of junctions between the structural elements make it challenging to predict the flanking transmission of noise [23,27,28].

It should be noted that at various conditions, the separation wall could be of high standard AIC, while the side wall could be of low standard, which is continuous between the rooms. In such cases, flanking transmission can propagate more noise than direct transmission [21]. Hence, the joints between the separation elements should be properly detailed to minimize noise due to flanking transmission.

3. AIC Properties

The ability to reduce sound transmission through a structural element is what defines the properties of AIC. Usually, concrete is a good insulator of sound, which can reflect up to 99% of its total energy. However, it is also a relatively poor absorber of sound and can propagate within enclosed spaces, causing echoes [29–32]. Sound, a form of energy, can travel through mediums of solid, liquid, and gas in longitudinal wave by the oscillation of vibrating particles [1,33]. As the sound waves dissipate, it gets expanded outward while distributing the intensity over a larger area. The larger the vibrating particles in the medium, the more the energy that passes through the medium [34]. Meanwhile, audible sounds are classified into two types: airborne and impact sounds. Airborne sounds can be recognized as speech, loudspeakers, and musical instruments, respectively. These sounds waves travels through the air only and cannot travel through solids [3]. The airborne sound produces vibration in the concrete, which causes the vibration of suspended particles on the other side of the wall, thus, causing it to be heard. On the other hand, footsteps, closing doors, and falling objects are categorized as impact sounds, which can vibrate between walls and floors, leading to airborne noise in adjoining rooms [2]. Typically, noise and sound differ from each other, with the former being subjective and dependent on the receptor. Due to it being subjective, architects and designers should consider reducing noise in the building, in urban areas in particular [35]. However, since it is challenging to lower the volume or the propagation of sound, noise mitigation measures should be implemented to bring down the level of annoyance [36,37]. In this regard, remedial measures such as insulation, reflection, and isolation provide the best ways to reflect sound energy [3,26].

4. Types of Acoustic Insulating Materials

The common aspects concerning the conservation of acoustic energy according to the classic ratio (1) are shown in Figure 3.

$$E_i = E_r + E_e \tag{1}$$

where E_i—energy of sound falling on the building envelope; E_r—reflected sound energy; and E_e—comprises the absorption and transmission of sound energy.

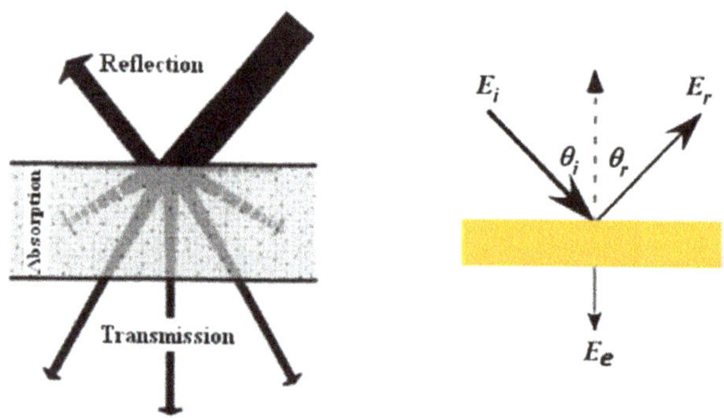

Figure 3. Sound energy conservation [14]. Reprinted with permission from Elsevier [14].

For the design of enclosing structures, it is important to use extremely sound-absorbing or extremely sound-reflecting materials.

4.1. Sound-Absorbing

The acoustic absorption coefficient α is computed using Equation (1) is expressed as (2).

$$\alpha = 1 - \frac{E_r}{E_i} \tag{2}$$

The absorption coefficient varies between 0 and 1, where 0 denotes the materials that reflect sound, and 1 is an ideal sound absorption material [38–40]. The presentation of the acoustic absorption coefficient as one value is an intricate problem.

This is attributable to the fact that the coefficient of acoustic absorption, α, is varied for each frequency of sound. As a simplification, the division of the audible frequency spectrum into octaves (with a further 1/3 octave division) is used, as well as, for example, such characteristics as noise reduction coefficient (NRC) and sound absorption average (SAA) [8,16,41]. NRC is a part of the easiest approaches that could be estimated as the arithmetic mean of sound absorption coefficient at frequency ranging between 250 and 2000 Hz. Meanwhile, SAA is the arithmetic mean of the coefficients of sound absorption at 12 one-third octave frequency ranging 200–2500 Hz in conformity with ASTM C423-17 [42]. Construction materials are commonly ranked by their NRC [42,43] (Figure 4).

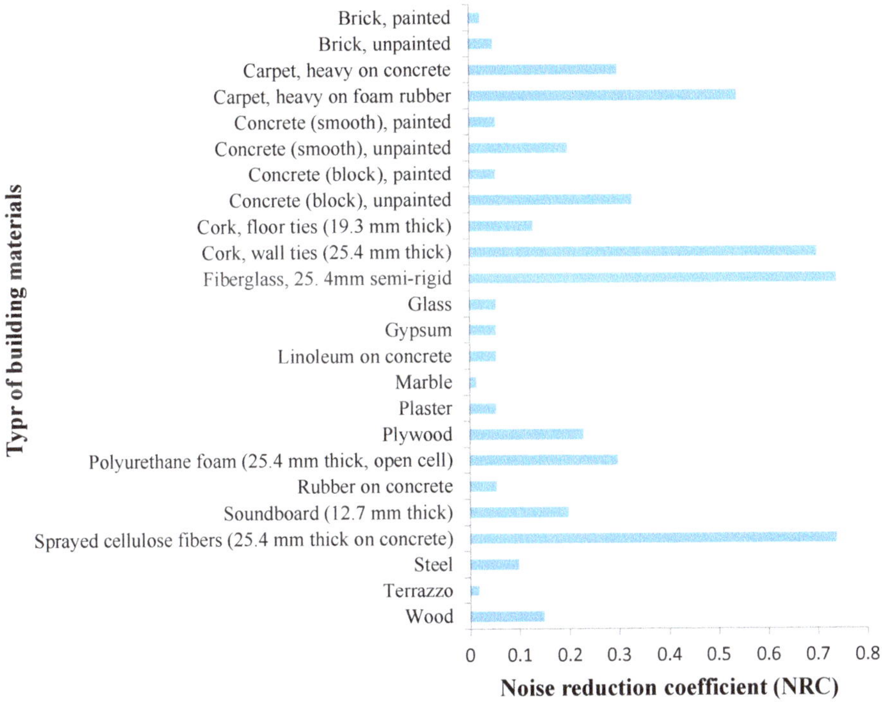

Figure 4. Building materials versus noise reduction coefficient (NRC) [43]. Modified with improvements from [43].

Summary of the noise reduction coefficients for different concrete composites are presented in Table 1. It can be concluded that an acoustic insulation concrete (AIC) building structure should possess a relatively good absorption coefficient with the NRC exceeding 0.45.

Table 1. A summary of NRC for different concrete composites.

Ref.	Year	Type of Specimen	Descriptions			Density (kg/m³)	α at Octave Frequency (Hz)				NRC
			Percentage (%)	Type of Materials	Size of LWA, mm (Thic. of Specimen, mm)		250	500	1000	2000	
[44]	2004	Concrete with LWA	20%	Lightweight cenospheres	10 to 300 μm (25 mm)	2310	0.04	0.10	0.21	0.17	0.13
			40%			2250	0.05	0.11	0.23	0.21	0.15
			60%			2180	0.05	0.10	0.11	0.12	0.11
[45]	2009		10%	CR	<4.75	2170	0.12	0.12	0.32	0.17	0.18
			20%			2110	0.11	0.10	0.37	0.15	0.18
[40]	2010		-	Expanded shale with 1.0% AE agent	4–8	1254	0.08	0.24	1.00	–	0.44
			-	Expanded shale with 1.0% AE agent	8–12	1189	0.06	0.22	0.96	–	0.41
[46]			10%	CR	<4.75	2200	0.05	0.10	0.06	0.07	0.07
			20%			2100	0.07	0.08	0.09	0.10	0.09
			30%			2000	0.12	0.13	0.11	–	0.12
[47]	2011	Hemp concrete	10, 15, 75%	Lime		1060	0.4	0.48	–	–	0.44
[48]	2012	Concrete with LWA	60%	Bottom ash	<10	701	0.21	0.24	0.28	0.28	0.25
[49]	2013	Mortar with LWA	80%		1–10 μm	1470	0.65	0.62	0.61	0.56	0.61
[50]	2014	Plaster with lightweight	80%	Insulation plaster and aerogel	– (10 mm)	300	0.03	0.08	0.06	–	0.06
[51]	2015	Alkali-activated cellular concrete	5%	Foam dosage	-	1050	0.05	0.10	0.15	–	0.10
			10%		-	960	0.06	0.10	0.20	–	0.12
[52]		Pervious concrete	1: 5	Vermiculite	0.5 to 4	640	0.10	0.21	0.78	0.32	0.35
[53]	2016	Hemp concrete	0.5% and 30%	30% GGBS and 0.5% ethyl cellulose	-	522	–	0.52	0.45	0.53	0.50
			30%	GGBS	-	505	–	0.49	0.42	0.44	0.45
			80%, 20%, and 0.5%	Hydrated lime, MK, and methyl cellulose	-	469	–	0.42	0.37	0.41	0.40
			80% and 20%	Hydrated lime, MK	-	493	–	0.46	0.39	0.44	0.43

Table 1. *Cont.*

Ref.	Year	Type of Specimen	Descriptions			Density (kg/m³)	α at Octave Frequency (Hz)				NRC
			Percentage (%)	Type of Materials	Size of LWA, mm (Thic. of Specimen, mm)		250	500	1000	2000	
[54]		Concrete with LWA	20%	CR	4–8	2264	0.06	–	–	–	0.05
			40%			2156	–	0.45	–	–	0.04
			60%			2026	–	0.54	–	–	0.06
			80%			1858	0.06	–	–	–	0.05
			20%	FCR		2313	–	0.30	–	–	0.05
			40%			2139	–	–	0.34	–	0.06
			60%			2032	–	–	0.43	–	0.10
			80%			1851	–	–	–	0.23	0.20
[55]	2017		10%	Miscanthus fibers	2–4	1504	0.06	0.20	0.25	0.07	0.15
			20%			1406	0.02	0.06	0.36	0.00	0.11
[illegible]		Geopolymer concrete with LWA	80%	C and D waste	<10	1510	0.00	0.06	0.85	0.23	0.29
[56]		Hemp concrete	-	Hemp shiv	5	590	0.13	0.28	0.91	0.48	0.45
[57]		Pervious concrete	50%	Arlite		700	0.08	0.12	0.46	0.23	0.22
[58]	2018	Alkali-activated cellular concrete	20–50%	Fly ash and 3:1 by mass. Adequate foam	-	940	–	0.25	–	–	0.19
						1130	–	0.23	–	–	0.24
						1310	–	0.18	–	–	0.11
[59]			35%	Foam dosage	-	600	0.20	0.10	0.40	0.94	0.41
			30%		-	720	0.18	0.16	0.54	0.78	0.42
			25%		-	820	0.03	0.12	0.43	0.85	0.36
[60]	2019	Concrete with LWA	50%	Polystyrene granules	1–4	1810	0.16	–	–	–	0.18
			50%	Polyethylene terephthalate	1–4	2047	0.22	–	–	–	0.18
			50%	Corn cob granules	size 1–6	1775	0.20	–	–	–	0.19
[61]		Hemp concrete	1:2	Hemp shiv	-	605	0.13	0.31	0.81	0.48	0.43
			1:2	Hemp fiber	-	407	0.19	0.63	0.83	0.71	0.59
[62]	2020	GGBS-based concrete	5–30%	GGBS as coarse and fine aggregates	1–4 and 4–8	419–995	-	0.54	-	-	0.24
[61]		Foam-glass concrete	92%	Foam bubbles	0.5–1.35	107–143	-	-	0.57	0.67	0.56

4.2. Sound-Insulating

Sound reflectors are materials that reflect noise, preventing the further propagation of a sound wave. Concrete, glass, wood, plastic, and metal are used as sound-reflecting materials.

In the concept of "Sound reduction index", two quantities should be distinguished [47]:

- R_w—sound insulation excluding flanking structures such as longitudinal walls, ceilings, floors, i.e., it is a value obtained during research in a laboratory;
- R'_w—sound insulation of the structure, taking into account the flanking elements, i.e., sound transmission along longitudinal side walls, floor slabs.

The last parameter is more adequate; therefore, it is the one that is important for assessing sound insulation. It varies widely for different types of concrete. For example, rubber concrete with a smooth surface has this value at the level of 30–35 dB [54], and for concrete with crumb rubber, it is 30–50 dB [3]. Moreover, for hemp concrete, it fluctuates in the range of 3–11 dB [56], and for homogeneous concrete floors, this value reaches 60–70 dB, and theoretically, it can reach 90 dB [63].

The same indicator is "Transmission Loss (TL)", which differs depending on the type of material. However, the sound reduction index R is defined by ISO standards, so, in general, it is usually preferred in the American literature. In addition, ISO 717-1 distinguishes between frequency-dependent R (or R′) and weighted values (R_w or R'_w). The latter values are accepted for frequencies around 500 Hz; therefore, they are considered the most presentable, and the fact is that a significant part of the sound (noise) spectrum is concentrated in the region of 500 Hz. Accordingly, ISO 717-1 requires the weighted Rw (or R′w) value precisely in the frequency range of about 500 Hz. Thus, this is a somewhat simplified, but sufficient for practical application, interpretation for sound reduction index R (or R′). For the sake of clarity, soundproof building materials are commonly designed in such a way as to have a maximum sound absorption coefficient, specifically for noise with frequencies in the region of 500 Hz.

The efficiency of sound reflectors is evaluated by the following features of materials and structures:

- The surface weight of the cladding [64,65]. In building acoustics, there is a well-known "law of mass", according to which doubling the mass of a single-layer fence leads to an increase in sound insulation. This works even more efficiently if the mass is concentrated in the surface layer, as can be seen from [66]. In addition, with an increase in the mass of the cladding, the resonant frequency of the system decreases, this also increases the sound insulation.
- The tightness of the structure. Slots and holes noticeably reduce the sound-insulating ability of the structure due to sound diffraction. For example, if a through hole of 2×2 cm in size is made in a 15 m^2 partition, then the sound insulation of the partition will decrease by 20 dB [67].
- The presence of a sound absorber inside the frame allows for a multilevel dissipation of sound energy [68]. Thanks to such measures, resonances in the airspace become impossible.
- The depth of the cladding frame. With the distance of the cladding from the wall, the sound insulation grows. This is due to the fact that the resonance frequency of the structure with which the claddings begin to effectively perform sound-reflecting functions is reduced. For example, when doubling the air gap of the tested structure, the increase in sound insulation is reported without increasing the cost of the structure [69,70].
- The absence or minimization of rigid connections, for example, places of fixtures, which are the bridges of sound [34,71].

5. Type of Concretes

Table 2 shows acoustic property of different type of concretes. It can be seen that normal concrete detects the highest level of sound reflection followed by glass-based concrete.

Table 2. Acoustic property of different type of concretes.

Type of Concretes	Maximum Coefficient of Sound Absorption	Level of Sound Reflection	Maximum Decrease in Sound Level at Frequencies, Hz	Refs.
Normal concrete	0.05–0.10	High	3000–5500	[52,57]
Aerated concrete	0.15–0.75	Low	250–2500	[44,72]
Foamed concrete	0.13–0.50	Low	100–2000	[51,61]
Crumb rubber concrete	0.30–0.70	Medium	400–2500	[3,54]
Polyurethane concrete	0.08–1.0	Low	150–1400	[55,60]
Coal bottom ash concrete	0.05–0.31	Medium	500–3500	[41,58]
Coconut fibers concrete	0.42–0.80	Medium	1250–3200	[59]
Recycled aggregate concrete	0.01–1.0	Medium	1500–2000	[60]
Oyster shell waste aggregate concrete	0.43–0.53	Low	1000–1800	[73]
Polymer concrete	0.90–1.0	Low	64–1600	[55]
Glass-based concrete	0.20–0.37	High	250–3150	[61]

5.1. Normal Concrete

The acoustic absorption coefficient for normal concrete is almost about 0.02, signifying that around 98% of the sound dynamism is a surface reflection. The more denser/heavy concrete, the higher the sound insulation rate that can be detected [3,18].

5.2. Aerated Concrete

In aerated concrete, alumina reacts with calcium hydroxide and leads to the appearance of gaseous hydrogen, which later forms microscopic bubbles in the mix. Lightweight aerated concrete has a density typically ranging 250 to 500 kg/m^3 and has a higher porosity of about 82.1% to 91.5% [72]. The pore wall of this concrete is very thin due to which sound waves can be transmitted very easily. According to Laukaitis and Fiks [74], the porosity decreases as the density increases, but on the other hand, the volume of open pores also increases. The open pore volume of aerated concrete is more than that of foam concrete. The absorption coefficient of autoclaved aerated concrete is best categorized by the ratio of open pores to total porosity.

5.3. Foam and Porous Concretes

Zhang et al. [51] developed a synthesis geopolymeric foamed concrete (FC) with absorption coefficients of 0.7 to 1.0 at 40–150 Hz and 0.1 to 0.3 at 800–1600 Hz, and 12 MPa of compressive strength was almost attained. Luna-Galliano et al. [75] studied the coefficient of absorption curves of geopolymers with silica fume used as a porous former. For different proportions of the starting components, the coefficient of sound absorption curves was the same, including with two peaks at 400 and 2500 Hz. It was noted that the mix contained more silica fume, which leads to the curve becoming wider (which indicates the stability of the coefficient of sound absorption), which can correlate with the highest open porosity. It was also found that an FC with low density exhibits a higher sound absorption coefficient (α) than a high-density FC when determined using a standing wave apparatus [76–78]. In Figure 5 [79], it can be observed that at the target void ratio (TVR) of 20%, the absorption coefficient revealed the highest value in the frequency range of 315–400 Hz; for the TVR of 25%, it ranked the second highest in the frequency range of 400–500 Hz; and for the TVR of 30%, it became the highest in the frequency range of 500–630 Hz [79]. This indicates that the frequency of the highest absorption coefficient was found for sound as the TVR augmented.

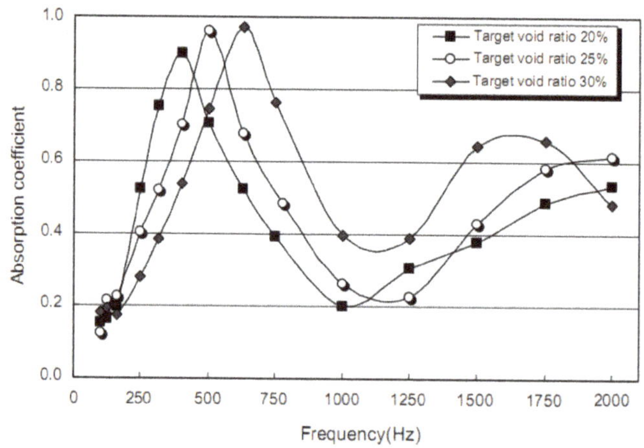

Figure 5. Sound absorption coefficient for the porous concrete with different densities and target void ratios (TVRs) [79]. Reprinted with permission from Elsevier [79].

In another study, it was found that the inclusions of 30% slag and fly ash into geopolymer FC (GFC) show no significant effect on the absorption coefficient at lower frequency but increases at higher frequency, 800–1600 Hz in particular [51,80]. This can be explained by the variation in the pore size, porosity, and tortuosity owing to inclusions of the slag particles [81]. Even increasing the dosage of foam (5 to 10%) shows little or no effect on the absorption coefficient at lower frequency but is efficient at medium frequency only, i.e., about 600–1000 Hz. Nevertheless, in comparison to a conventional plain concrete, which has an absorption coefficient typically less than 0.1 at 125–2000 Hz, GFC shows excellent acoustic absorption properties [82]. However, GFC has low sound absorption at medium to high frequencies relative to expanded clay porous concrete, which usually has an absorption coefficient of about 0.5 in this region [83]. However, a GFC shows the same absorption coefficient as the thickness of the GFC increases (Figure 6).

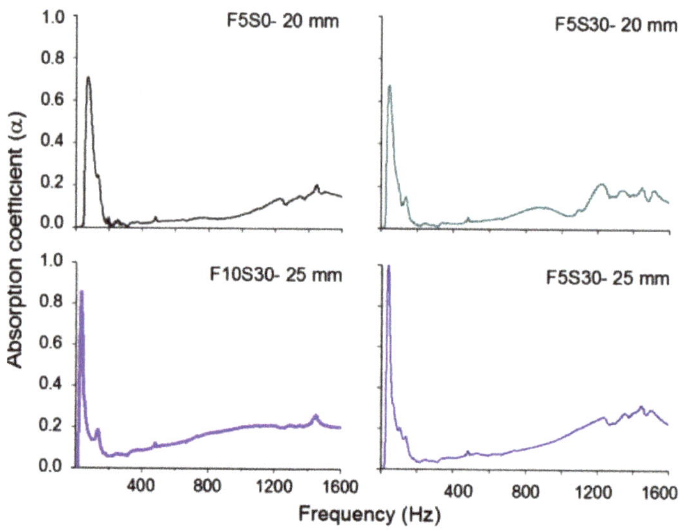

Figure 6. Acoustic absorption spectra of geopolymer foamed concrete (GFCs) at varying thickness [51]. Reprinted with permission from Elsevier [51].

5.4. Crumb Rubber Concrete

Crumb rubber concrete (CRC), which is produced using different sizes of crumb rubber typically ranging 6 to 19 mm for coarse and 1 to 6 mm for fine, has been a subject of recent research [3,31,32,62,84]. However, the use of CRC in structural applications is not practical owing to considerable reduction in the strength [85,86]. Although CRC has lower mechanical strength, it can be used as a durable composite material capable of reflecting and absorbing sound [87]. In fact, the use of CRC as an exterior surface of a building can reduce the propagation of street noises into the buildings. High noise levels caused by busy streets and passing into dwelling spaces are often uncomfortable for inhabitants in high-rise apartments. An extremely dense material has an absorption coefficient of about 0 and can reflect away the sound completely. Hence, the use of CRC can provide a solution to both sound absorption and recycling used tires. Table 3 presents a list of classic absorption coefficients for some common materials used in structural concretes.

Table 3. Sound absorption coefficients for some common materials.

Type of Material	SACs
Concrete	0.02–0.06
Hardwood	0.3
Unpainted block-work	0.02–0.05

Materials with good sound absorption prepared by a concrete matrix incorporated with crumb rubber were investigated in [56]. The sound absorption coefficient α was above 0.5 for nine out of 12 specimens and attained a maximum value of 0.82 and 0.93 under favorable conditions. The concrete panels were prepared with crumb rubber as a replacement to natural fibers and investigated its acoustic properties in references [3,88]. The results showed that rubber concrete has proven itself in terms of sound absorption, especially with a high percentage (15%) of rubber crumb. Sukontasukkul [45] reported an improvement of 46% in the NRC upon the inclusion of 20% crumb rubber in precast concrete panels.

Materials containing crumb rubber have been known to be a good absorber of sound, with absorption coefficients ranging between 0.3 and 0.7 [32]. In fact, the combination of crumb rubber and concrete further can increase the absorption coefficient as well as reduce the level of reflected sound [89]. Apart from structural concrete, the use of crumb rubber has also been known to be a good sound absorber when utilized in highways. The incorporation of crumb rubber in asphalt mixes could lower the noise produced by vehicular traffic. The sound absorption coefficients of asphalt pavements containing crumb rubber are considerably improved with time due to its higher absorption energy [46]. The damping vibration was about 230% higher when crumb rubber was used to replace 15% of natural fines in CRC [90]. Incorporating 20% crumb rubber in concrete blocks was also reported to produce a lighter, flexible, and durable sound-absorbing material [91]. The durability of CRC against freeze–thaw could also be improved using air entraining admixtures [92,93]. It is found that CRC when exposed to excessive heat shows a substantial enhancement in the energy absorption, but only when lower grades of rubber and natural fines substitution were considered [85]. In comparison to conventional aggregates, crumb rubber is highly elastic, which deforms easily under an applied load, and its ductility can increase up to 90%, subsequently improving energy dissipation [3]. As the crumb rubber content increases, the density of CRC reduces. Since a material's acoustic properties depend mostly on its density, lightweight concrete such as CRC containing a higher proportion of crumb rubber could absorb more sound. Hence, the ability to reflect more sound energy increases [32,94]. Due to this, CRC panels have been used mostly in office buildings as exterior cladding or on balconies of buildings. The compaction degree of a CRC also greatly influences its sound absorption, since the larger grades of CRC absorb more sound when fully compacted [62,90]. Meanwhile, CRC cladding panes are often used as an alternative to the conventional ones to protect the structure with an added benefit of reducing the overall weight. A proposal provided by the City of Vancouver Noise Control Manual [95] shows that the installation of exterior CRC panels around the balconies of high-rise buildings can improve its acoustic performance with a reduction in the level of reflected noise as well (as see in Figure 7).

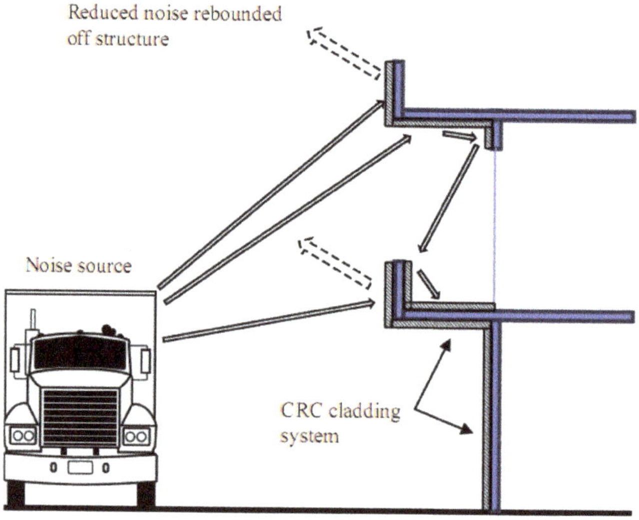

Figure 7. A crumb rubber concrete (CRC) cladding installed on a multi-storey structure to avoid the diffusion of sound [95]. Modified with improvement from *Wakefield* [95].

In reference [96], the importance of preliminary processing of the surface of secondary rubber particles to increase the sound-insulating properties of CRC was proved. The elastic modulus of concrete combinations was determined, and the surface adhesion prepared from pretreated/untreated rubber aggregate was studied [3]. Conclusions indicated that the method of pretreatment caused weaker adhesion between the cement paste and crumb rubber; therefore, it improved the ability to absorb the vibration and sound-insulating properties of concrete. It was shown that the effects of freezing and thawing do not significantly affect the insulating properties [3]. It was been found that the insulation characteristics for all concrete improved at high frequencies. The findings show that rubber concrete can be used on the outside of building envelopes to absorb sound throughout multi-storey urbanized buildings, but it requires full-scale on-site testing.

5.5. Expanded Polystyrene Concrete

The use of foamed polystyrene aggregates has some effect on noise reduction owing to its lower stiffness, which has a high ability to dissipate energy. Expanded polystyrene granules (EPS) can easily be added in concrete to produce a lightweight concrete of varying density [60]. EPS are ultra-lightweight aggregates with a density <33 kg/m^3, non-absorbent, hydrophobic, and closed cell nature [97]. The NRC of a 50 mm-thick layer of EPS is around 0.32 [98]. This means that about 32% of the incident sound energy gets absorbed and does not reflect back, hence improving the absorption characteristics of the material. Oancea et al. [60] replaced the aggregate with 50% polystyrene foam while obtaining lightweight concrete with a density of 1810 kg / m^3 and NRC at the level of 0.18. Due to the closed porosity of polystyrene beads, effective damping of the sound wave is not possible.

5.6. Fibered Concrete

At present, synthetic fibers such as glass and mineral fibers are used in buildings as a sound-proofing material [99]; however, these pose risks to human health [100]. It is also found that the airflow resistivity is an important parameter that needs to be considered in the acoustic behavior of fibrous concrete [12]. When choosing raw materials for concrete production, it is necessary to select materials with a reduced density that is capable of damping the sound wave. In particular, when choosing fiber, polypropylene has one of the lowest densities (Figure 8).

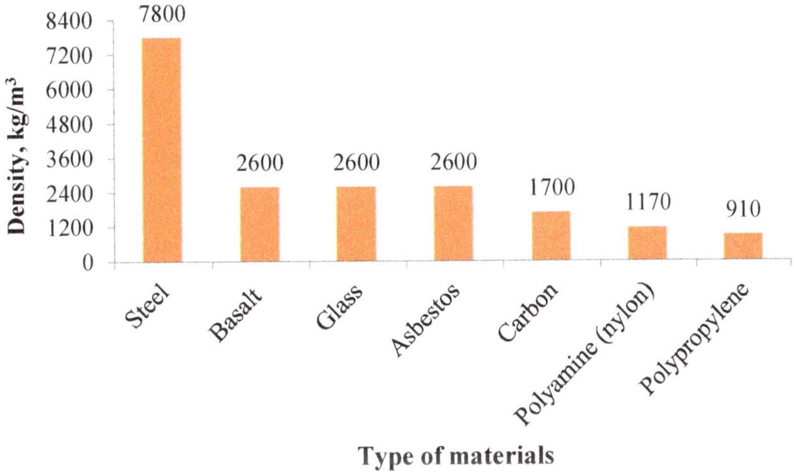

Figure 8. Densities of materials of various fibers [101,102]. Data adopted with permission from Fediuk and Klyuev [101,102].

Stainless steel, nickel fiber, alloy fiber, and iron metallic fiber are some of the noise reduction materials available [12]. A review of existing literature on the acoustic absorption of metallic and inorganic fibrous composites is presented in Table 4. The use of fibers is widely known to provide a wide and varied solution to noise reduction in buildings, some of which are listed in Table 4.

5.7. *Recycled Aggregate Concrete (RCA)*

The sound absorption of a cellular material is mostly influenced by the energy loss occurred due to the friction induced within the pores of the walls [79]. Hence, the higher the open void ratio, the greater the SAC of concrete at all frequencies [103]. In a porous concrete, the pores are interconnected to each other, and the pores increase as the amount of coarse aggregate increases [104]. Consequently, the sound absorption of a porous materials is significantly influenced, which is directly proportional to its airflow resistivity [105]. Hence, to produce a sound-absorbing concrete, it is imperative to generate a tortuous path wherein the sound energies could dissipate, mostly in frequencies of noise traffic, ranging between 300 and 1000 Hz [106]. Some researchers used recycled aggregate with particle sizes greater than 5 mm as a substitute for aggregates in concretes [13,107].

Table 4. Acoustic insulation of different fibrous concrete composites.

Type of Fiber	Main Findings	Refs.
Rock wool	Similar acoustic behavior to glass wool	[108]
Carbon and glass fiber	Composites made with carbon fiber has higher SAC relative to glass fibered composite	[109]
Fibrous metal materials	Used to make silencers in cars	[110]
Glass wool	Comparison made between the Bies–Allard and Kino–Allard's acoustic methods	[111]
Metal fiber felts	Used as an absorption material in silencers	[112]
Glass fiber-reinforced epoxy	Investigated the acoustic absorption properties of different composites	[113]
Sintered fibrous metals	Determined anisotropic acoustic properties of of sintered fibrous metals	[35]
Glass fiber recycled from deserted print circuit boards	Utilized for noise-reducing applications	[114]
Metal fiber	Absorption properties depends on the material properties such as the diameter, porosity, and thickness of fiber	[115]
Glass fiber felt	The direction of sound incidence and structure of the composite affects the sound insulation	[115]
Carbon fiber	Increases the sound absorption coefficient of a helical-shaped composite sound absorber	[116]
Fouled sintered fiber felts	Depends on the flow resistivity measurements	[117]
Basalt fiber	Panels shows a good absorption coefficient that increases with thickness and density	[118]
Carbon fiber	Composites made with carbon fibers shows higher absorption coefficients than Kevlar fiber at low to medium frequencies	[119]
Glass fiber-filled honeycomb sandwich panels	Improves the absorption coefficient at frequencies below 4.5 kHz	[120]
Metal fiber porous materials	Porous material can effectively enhance the sound absorption coefficient	[121]

At the same time, the SAC was found to be enhanced with the use of recycled concrete aggregates [11]. An SAC of about 0.91 was recorded when 100% of RCA was used, which is 0.24 higher than the control mix [122]. This was as a result of the greater void content in the RCA concrete. Figure 9 [13] clearly portrays that the SAC increases as the demolition wastes aggregates (CDW)/fly ash (FA) increases. In fact, the porous concrete corresponding to the CDW80-FA20 mix showed the best SAC. Meanwhile, a huge variation can be observed when the amount of coarse fraction increases to 80%.

5.8. Mollusk Shell Waste Aggregate Concrete

It is also known that the particle size plays a crucial role in the acoustic properties of porous materials made with shell waste [49]. To examine this, the grain size distribution of the mollusk shell waste (7 and <−2 2 mm) was studied [41]. It was found that the porous concrete containing shell waste aggregates of about 5–10 mm showed the best acoustic absorption properties [49,79]. Meanwhile, it is reported that the internal pores have no contribution to the SAC of a porous concrete [55]. Hence, it was concluded that concretes made with mollusk shell waste (2 to 7 mm particle size) increase the SAC by about 40% with respect to the porous concrete containing natural aggregates only. [123]. For particles of the same size, the sound transmission lost showed absorption of more than 4 dB at a thickness of 12 cm. An improvement in the acoustic characteristics of the composites is possible due to the pyrolysis of mollusk shells, resulting in a convoluted surface capable of effectively damping the sound wave [73]. This finding coincides with the study by Paridah et al. [124] wherein the sound waves hitting the pores of a wall produce an acoustic wave due to the dissipation of energy. Due to the porosity and tortuosity in the specimen, an incident sound wave is produced, dissipated into sound energy due to internal friction, and results in high SAC [125]. Logically speaking, large particles cause a reduction in the energy dissipation due to its higher porosity, and hence, the absorption coefficient also decreases. Hence, sizes up to 7 mm are crushed and allowed to be utilized [41,49]. Owing to this, the SAC of a porous concrete is usually lower than a recycled concrete, mostly in frequencies below 2000 Hz [123].

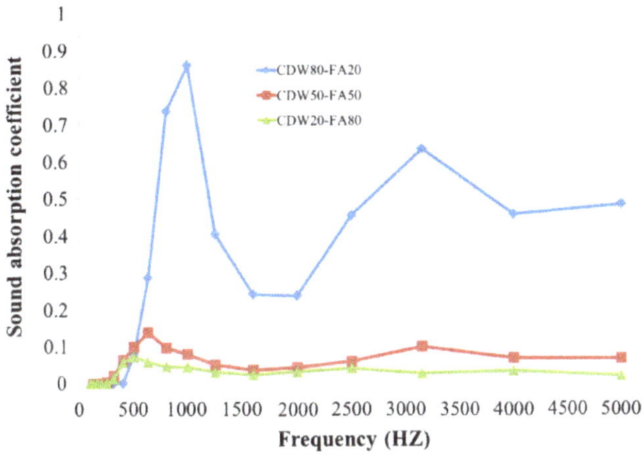

Figure 9. Sound absorption coefficient (SAC) for demolition wastes as aggregates for geopolymer concrete [13]. Reprinted with permission from Elsevier [13].

5.9. Polymer Concrete

Arenas et al. [13] report that the activation solution in the geopolymerization of fly ash has a considerable effect mainly on the strength parameters but not on acoustic properties. However, according to Luna et al. [75], incorporating a higher amount of silica fume in the mix and increasing the setting temperature leads to an increase in the open voids. In turn, this reduces the compressive strength but increases the SAC. Similarly, Mastali et al. [59] developed cinderblock alkali-activated concrete with significant acoustic characteristics for the interior walls of buildings in residential areas. Stolz et al. [58] produced geopolymers with a density of about 1000 kg/m^3 as well as heat and sound insulation comparable to the properties of the commercial existing options. In comparison to ordinary Portland cement (OPC)-based composites, alkali-activated FC shows excellent acoustic absorption characteristics [44]. Porous concrete was also found to exhibit average SACs higher than 0.5, at mid to high frequencies [83]. The alkali-activated FC was also observed to exhibit good SAC (>0.5) at the same frequency levels as that of the porous concrete [59].

5.10. Foam-Glass Based Concrete

The use of foam-glass concrete (FGC) can also provide a good acoustic insulation wherein the annealing step is not required, and the density can also be greater than 150 kg/m^3 [126]. In fact, foamed-glass blocks of high costs can exhibit better aesthetics for indoor acoustic insulation properties [127]. However, an FGC composite is a unique material with good acoustic properties, with an almost unlimited service life [126,128–131]. The main reason for the insufficiently widespread use of foam glass in structures is not its operational features, but the cost related with the technological features of its manufacturing. Glass fiber is widely used in cement materials, which do not show a considerable improvement in sound absorption, but it improves sound reflection well due to the creation of a dense and low-porous structure [128,132–135]. Although significant studies have investigated FGC, very few characterized its open and closed porosity in detail [136–138]. The majority of the available research focused only on the partially closed, open porosity, closed porous with high density, or open porosity with lower-density FCs [139–141]. FGC with open porosity is usually due to the use of a decomposition foaming agent that crystallizes [139–144]. This type of concrete is produced using a replicate synthesis that was obtained from organic foams, the sol–gel process, or inorganic gel casting [127,145,146]. Meanwhile, only a few studies are available on FGCs with closed porosity and a density lower than 150 kg m^{-3} [139,147–152]. The SAC as a function of frequency containing 20% glass and 80% GDL (sample 6 and 10) has also been studied (Figure 10) [127]. From this study, it has been found that findings were insignificant at frequencies lower than 100 Hz. Meanwhile, a few researchers observed that the SAC of glass foam is higher than rock wool at medium to high frequencies [153–155]. This is mainly due to the porous nature of the material. Since foam is composed of an open cell structure, it shows excellent performance at a frequency range of 1250–3150 Hz [156]. Similarly, tortuosity also causes the SAC to vary while in transit from an open cell structure to a fibrous one [157]. In addition, due to the dimensional stability of FGC with open porosity, it is suitable as immovable acoustic insulation elements or as a provision for catalysts [145].

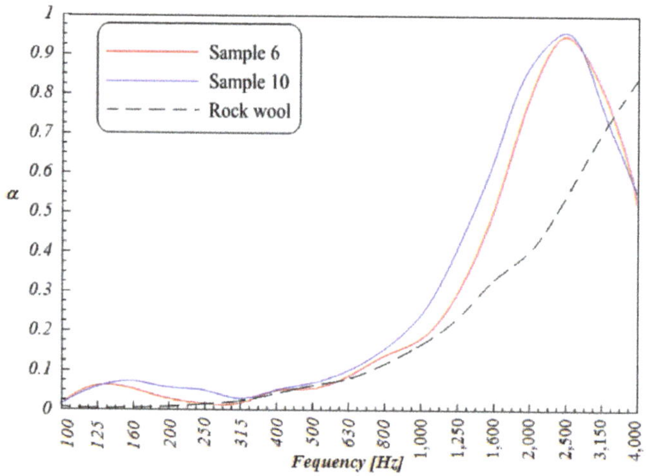

Figure 10. Average SACs of samples compared with rock wool [127]. Reprinted with permission from Elsevier [127].

6. Building Components

Evaluation of the acoustic characteristics of the building elements allows you to deal with the noise penetrating the premises and determine the degree of its impact [158,159]. Building acoustics study the transmission of sound (for example, the steps of people and the noise of passing cars) through the walls and entrances of a building [158,160]. First, sound is measured inside and outside the building, and then work is carried out in the room to change the difference in the levels of reverberation and background noise [161,162]. Information about noise (for example, about its frequency spectrum) allows you to effectively deal with it, for example, using isolation and shielding [163,164].

6.1. Reinforced Concrete Wall System

A reinforced concrete wall has better sound-insulating properties than a brick of the same thickness, because it is denser [96]. A thick concrete wall is better than a thin concrete wall in terms of sound insulation [19]. However, in order for a single-layer reinforced concrete wall to provide standard sound insulation for urban conditions, it is necessary to ensure that its thickness is more than 1 m, which is very expensive and significantly reduces the internal space [165]. Therefore, the use of multilayer wall systems for building facades is very relevant [166]. Figure 11 shows a clear model for the construction of RC walls with extremely high acoustic insulation. Carefully selected layer thicknesses and their sequence ensure a sound reduction index of up to 60 dB [167].

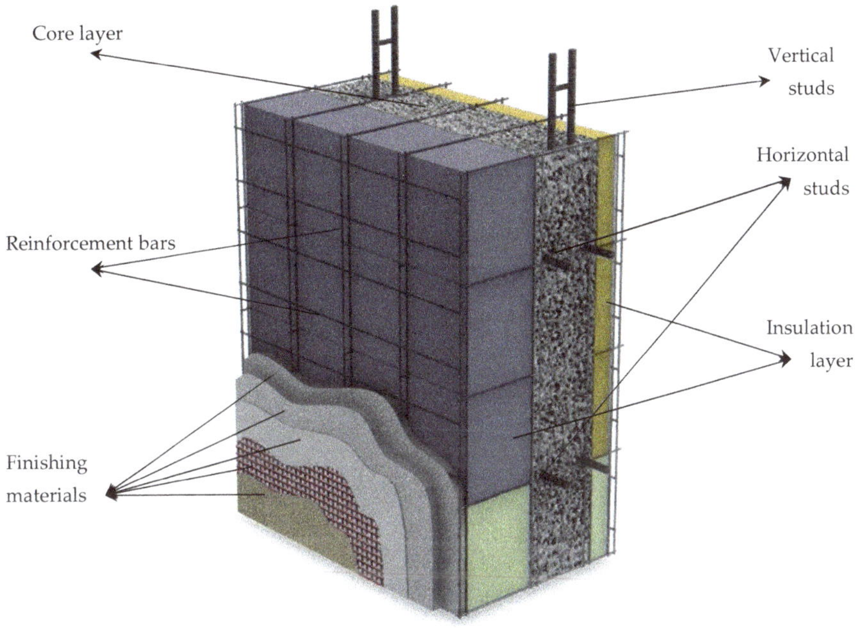

Figure 11. Model for the construction of reinforced concrete walls with extremely high acoustic insulation [167]. Reprinted with permission from ECOSISM [167].

6.2. Steel Plate Wall System

Good sound insulation properties are provided by wall systems using steel plates and an air inner layer, which are called ventilated facades [168]. The system of ventilated facades is not only an architectural solution, protection from adverse weather conditions, and reliable thermal insulation, but it is also an optimal sound-insulating property (Sound reduction index, R_w) that can be improved up to 15 dB [169]. One of the widely used approaches in reducing airborne noise transmission is by the installation of a denser acoustic mineral wool wall, as shown in Figure 12 [21]. This dense mineral wool slab shows an excellent SAC and also acts as an additional layer to the structural concrete.

The insulation layer is usually installed in addition to the insulation quick, which maintains the low-level coupling between the two skins. It also maintains the thermal properties of the structural concrete. Meanwhile, the usage of a mineral wool acoustic wall can compromise with the acoustic performance due to its higher degree of coupling [98]. However, the additional mass provided by the acoustic mineral wool wall can reduce the impact noise propagated through the structure during rains [21].

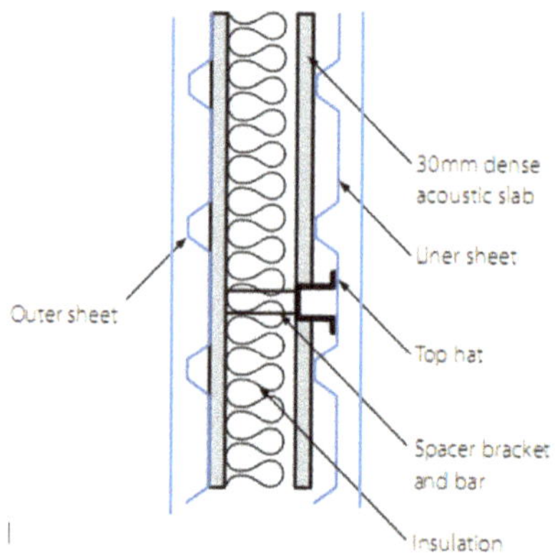

Figure 12. Effect of the of wall's thickness built-up on acoustic insulation [21]. Reprinted with permission from DATA STEEL [21].

6.3. Masonry Wall System

The compositional nature of the construction of the three-layer facade, combined with a brick wall, allows combining materials with excellent sound-absorbing and sound-reflecting properties (Figure 13) [170,171]. The mineral wool layer located inside has good sound absorption, which reduces the noise load by more than two times (TL and R_w), especially from low-frequency sounds [172,173].

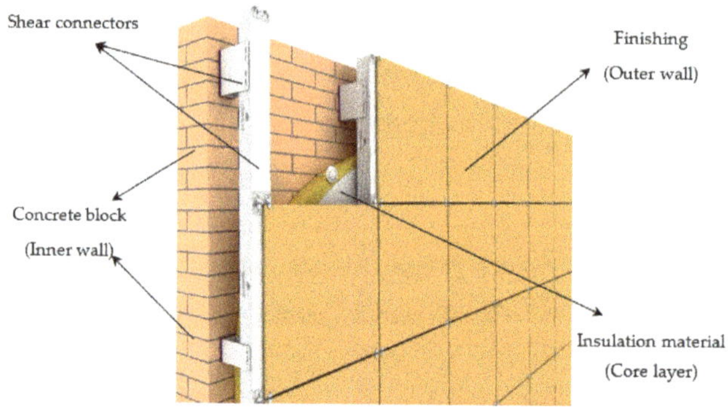

Figure 13. Masonry wall system [174]. Reprinted with permission from ORION-SD [174].

6.4. Concrete Sandwich Panels

A study by Frazao et al. [175] developed innovative structural panels, a combination of sisal fiber–cement composite as the thin outer layer, and a polypropylene-reinforced lightweight concrete as the base layer. Such multilayer composition makes it possible to efficiently perform sound-insulating functions by distributing sound-absorbing and sound-reflecting functions between the layers. Experimentally determined layer thicknesses and their sequence provide a sound reduction index R_w up to 50 dB [175]. Cuypers and Wastiels [176] made sandwich panels with outer layers of textile-reinforced concrete. Due to the fact that textile concrete is a light type of concrete, this allows for the creation of the effective sound absorption of a wave's incident on the wall, both outside and inside. In this case, the sound reduction index R_w of the developed design is 45 dB. Good acoustic properties (R_w = 40 dB) are predicted for multilayer panels with corrugated steel surfaces and a core of plain or reinforced foam, as developed by Flores-Johnson and Li [177].

6.5. Reinforced Concrete Slab System

In urban apartment buildings, ensuring the sound insulation of floors between floors is of great importance [63,178]. To do this, it is necessary to lower the level of impact noise through the usage of various damping additives in the ceiling, for example, rubber [179]. The transmission of impact sound through hard-walled ceilings is greatly increased (R_W = 45 dB); therefore, rigid joints must be avoided [180,181].

Traditionally, hollow concrete slabs are utilized to lower the volume of concrete in the slab and to minimize its own weight; however, in this case, the resistance to impact noise is reduced (R_W = 36 dB) [182]. This is due to the fact that, unlike wall structures, the sounds of steps, moved furniture, etc., which are effectively extinguished by massive structures, are transmitted through the ceilings [183]. Al-Rubaye et al. [184] developed a hollow composite reinforcing system with four flanges with an aim to enhance the adhesion of concrete, stabilize holes in concrete elements, and improve sound insulation performance (R_W = 43 dB). It is also reported that the built-up system of Cobute precast slab (Figure 14) had a 50% reduction on self-weight, dramatically assisted with services distribution, and effectively contributed acoustic insulation (R_W = 40 dB) [185].

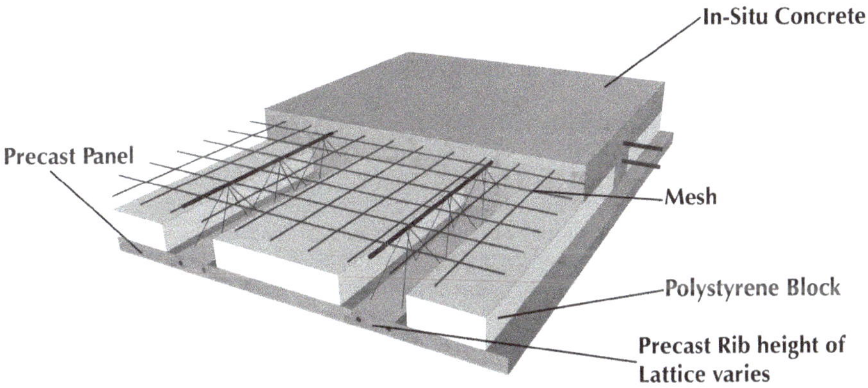

Figure 14. Details the Cobute precast slab system [185]. Reprinted with permission from Cobute [185].

6.6. Steel–Concrete Composite Floorings

In comparison to reinforced concrete slabs, composite floor systems offer a more economical solution (Figure 15) [186]. The constituent structural elements of the reinforced concrete slabs can optimize the strength and performances of the steel and the concrete as well. However, due to the connection of the rigid system "reinforced concrete slab–steel beam", impact and structural noise quite easily find their way to the lower floor. To improve the acoustic characteristics of the floors, various modifications of these composites have been developed, such as slim floor beams [187], Thor and Delta beams [188], composite slim floor beam [189,190], iTECH composite beams [191], and Ultra-Shallow Floor beams [192–194] are the other few composite flooring systems used. Some of these composite flooring systems use a profile steel sheet or small mesh beam [187,188,195]. This provides a better sound insulation to the floor (R_W = 30–50 dB).

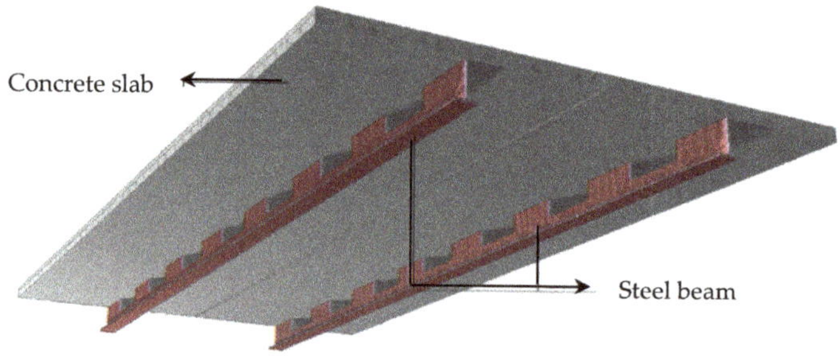

Figure 15. Steel–concrete composite floorings [186]. Reprinted with permission from Elsevier [186].

7. Conclusions

Under the conditions of the current world development, noise pollution associated with population growth and industrial development is the main problem faced by people living mainly in municipal areas. This condition highlights the importance for long-term research of new energy materials that can reduce the acoustic power of a sound wave as a result of absorption and reflection. From the literature reviewed, it is found that each construction material has a different NRC and fundamentally relies on its density. However, to avoid sound waves, in any building materials, as far as possible, mass is required—that is, high-density materials, such as bricks, are best suited for partitions, while cellular concrete is lightweight and inferior to bricks in sound insulation of wave noise. However, it is imperative to note that regarding the structural noise of lower frequency, a cellular concrete isolates better than brick. The construction industries worldwide have started to mainly use the sound-absorbing concrete to reduce the level of sounds in opened and closed areas and increase the amount of sound insulation.

To increase sound insulation, wall structures should be designed as multilayer, including with the use of air gaps. Slab floorings that work as absorbers of structural noise should be designed using damping components, such as chopped rubber, and it is also necessary to prevent rigid "sound bridges". It is proposed to expand the range of measured acoustic characteristics by expanding the lower boundary of the frequency range from 100 to 20 Hz.

However, it is obligatory to consider all structural parameters and design characteristics, when designing concrete structures, such as acoustic insulation property. This is, in particular, to ensure comfortable living conditions for people in residential premises, including acoustic comfort. Different types of concrete behave differently as sound conductors; especially dense mixtures are superior sound reflectors, and light ones are sound absorbers. It is found that the level of sound reflection in modified concrete highly relies on the type of aggregates, distribution and size of pores, and changes in concrete mix design constituents. The sound absorption of AIC can be improved by forming open pores in concrete matrices by either using a porous aggregate or foam agent. To this end, this article reviews the noise and sound transmission in buildings, types of acoustic insulating materials, and the acoustic properties. This literature study also provides a critical review on the type of concretes, the acoustic insulation of buildings and their components, and the assessment of sound insulation of structures, as well as synopsizes the research development trends to generate comprehensive insights into the potential applications of AIC as an applicable material to mitigate noise pollution for increase productivity, health, and well-being. Nevertheless, AIC is deemed as one of the superior materials that should be used in the building constructions, in particular, concert halls, cinemas, theatres, music venues, etc. Several further research investigations are recommended for the production of renewable and green concrete composite;

- New applications of AIC are worth exploring and can be found; for example, EPS-based concrete can be produced as a class of innovative lightweight soundproof concrete.

- To further study the potential use of modified concrete to develop high-sound insulation performance concrete.
- To increase the acoustic insulation performance of AIC in a hardened state using ecofriendly materials.
- To further extend the possible utilization of AIC in the building construction with a sound-insulated system and future sustainable cities with reduced noise and sound transmission.

Author Contributions: Conceptualization, R.F., and M.A.; methodology, M.A., R.F., and N.V.; validation, R.F., M.A., T.O., N.V., Y.V. and V.L.; resources, M.A., R.F., and N.V.; data curation, M.A., R.F., Y.V., T.O., and V.L.; writing—original draft preparation, M.A., and R.F.; writing—review and editing, M.A., R.F., N.V., Y.V., V.L. and T.O.; supervision, M.A., and R.F.; Visualization, N.V., Y.V., T.O., and V.L.; Revision, T.O.; project administration, R.F., M.A., and N.V.; funding acquisition, M.A., N.V., and Y.V. All authors have read and agreed to the published version of the manuscript.

Funding: This research is partially funded by the Ministry of Science and Higher Education of the Russian Federation as part of World-class Research Center program: Advanced Digital Technologies (contract No. 075-15-2020-934 dated 17.11.2020).

Institutional Review Board Statement: Not applicable.

Informed Consent Statement: Not applicable.

Data Availability Statement: Data sharing not applicable.

Acknowledgments: The authors gratefully acknowledge the financial support by Deanship of Scientific Research at Prince Sattam bin Abdulaziz University, Alkharj, Saudi Arabia, the Peter the Great Polytechnic University, Saint Petersburg, Russia, and the cooperation of the Department of Civil Engineering, Faculty of Engineering and IT, Amran University, Yemen, for this research.

Conflicts of Interest: The authors declare no conflict of interest.

Abbreviations

AIC	Acoustic insulation concrete
CR	Crumb rubber
CRC	Crumb rubber concrete
EPS	Polystyrene granules
FA/GGBS	Fly ash/Ground granulated blast furnace slag
FC	Foam concrete
FCR	Fiber crumb rubber
FGC	Foam glass concrete
GDL	Brown algae, D-Gluconic acid δ-Lactone
GFC	Geopolymer foam concrete
ID	Identifier
LWA	Lightweight aggregate
LWC	Lightweight concrete
NRC	Noise reduction coefficient
OPC	Ordinary Portland cement
RC	Reinforced concrete
Rw	Sound reduction index
SAA	Sound absorption average
SAC	Sound absorption coefficients
TL	Transmission loss

References

1. Patil, M.G.S.; Kulkarni, G.S. Analysis of Noise Pollution in Kolhapur City and Technical Remedy to Reduce Noise Level. *Int. J. Eng. Res.* **2020**, *1*, 219–282.
2. Héroux, M.E.; Braubach, M.; Dramac, D.; Korol, N.; Paunovic, E.; Zastenskaya, I. Summary of ongoing activities on environmental noise and health at the WHO regional office for Europe. *Gig. Sanit.* **2014**, *5*, 25–28.

3. Holmes, N.; Browne, A.; Montague, C. Acoustic properties of concrete panels with crumb rubber as a fine aggregate replacement. *Constr. Build. Mater.* **2014**, *73*, 195–204. [CrossRef]
4. Ibragimov, R.; Fediuk, R. Improving the early strength of concrete: Effect of mechanochemical activation of the cementitious suspension and using of various superplasticizers. *Constr. Build. Mater.* **2019**, *226*, 839–848. [CrossRef]
5. Brookhouser, P.E. Prevention of Noise-Induced Hearing Loss. *Prev. Med. (Baltim)* **1994**, *23*, 665–669. [CrossRef]
6. Cuthbertson, D.; Berardi, U.; Briens, C.; Berruti, F. Biochar from residual biomass as a concrete filler for improved thermal and acoustic properties. *Biomass Bioenergy* **2019**, *120*, 77–83. [CrossRef]
7. Kim, H.; Hong, J.; Pyo, S. Acoustic characteristics of sound absorbable high performance concrete. *Appl. Acoust.* **2018**, *138*, 171–178. [CrossRef]
8. Vinokur, R. Infrasonic sound pressure in dwellings at the Helmholtz resonance actuated by environmental noise and vibration. *Appl. Acoust.* **2004**, *65*, 143–151. [CrossRef]
9. Li, X.; Liu, Q.; Pei, S.; Song, L.; Zhang, X. Structure-borne noise of railway composite bridge: Numerical simulation and experimental validation. *J. Sound Vib.* **2015**, *353*, 378–394. [CrossRef]
10. Tsunekawa, S.; Kajikawa, Y.; Nohara, S.; Ariizumi, M.; Okada, A. Study on the perceptible level for infrasound. *J. Sound Vib.* **1987**, *112*, 15–22. [CrossRef]
11. Tie, T.S.; Mo, K.H.; Putra, A.; Loo, S.C.; Alengaram, U.J.; Ling, T.C. Sound absorption performance of modified concrete: A review. *J. Build. Eng.* **2020**, *30*, 101219. [CrossRef]
12. Tang, X.; Yan, X. Acoustic energy absorption properties of fibrous materials: A review. *Compos. Part A Appl. Sci. Manuf.* **2017**, *101*, 360–380. [CrossRef]
13. Arenas, C.; Luna-Galiano, Y.; Leiva, C.; Vilches, L.F.; Arroyo, F.; Villegas, R.; Fernández-Pereira, C. Development of a fly ash-based geopolymeric concrete with construction and demolition wastes as aggregates in acoustic barriers. *Constr. Build. Mater.* **2017**, *134*, 433–442. [CrossRef]
14. Palomar, I.; Barluenga, G.; Puentes, J. Lime-cement mortars for coating with improved thermal and acoustic performance. *Constr. Build. Mater.* **2015**, *75*, 306–314. [CrossRef]
15. Keränen, J.; Hakala, J.; Hongisto, V. The sound insulation of façades at frequencies 5–5000 Hz. *Build. Environ.* **2019**, *156*, 12–20. [CrossRef]
16. Møller, H.; Pedersen, C.S. Hearing at low and infrasonic frequencies. *Noise Heal.* **2004**, *6*, 37–57.
17. Hambric, S.A. Structural Acoustics Tutorial—Part 1: Vibrations in Structures. *Acoust. Today* **2006**, *2*, 21–33. [CrossRef]
18. Feduik, R. Reducing permeability of fiber concrete using composite binders. *Spec. Top. Rev. Porous Media* **2018**, *9*, 79–89. [CrossRef]
19. Gramez, A.; Boubenider, F. Acoustic comfort evaluation for a conference room: A case study. *Appl. Acoust.* **2017**, *118*, 39–49. [CrossRef]
20. Holger Rindel, J. *Sound Insulation in Buildings*; CRC Press: Boca Raton, FL, USA, 2017; ISBN 9781498700429.
21. Tata Steel Acoustic performance of pre-finished steel cladding systems. *Tata Steel UK Ltd.-2010* **2013**, *2*, 24.
22. Mommertz, E. *Acoustics and Sound Insulation*; Birkhauser Verlag AG: Basel, Switzerland, 2013.
23. Quartaruolo, G.; Beresford, T. Method to predict airborne flanking through concrete floors with nibs at the base of lightweight walls using ISO 12354-1. In Proceedings of the INTER-NOISE 2017—46th International Congress and Exposition on Noise Control Engineering: Taming Noise and Moving Quiet, Hong Kong, China, 27–30 August 2017.
24. Fediuk, R.S. Mechanical Activation of Construction Binder Materials by Various Mills. In Proceedings of the IOP Conference Series: Materials Science and Engineering, Yurga, Russia, 26–28 November 2015; Volume 125.
25. Park, H.S.; Oh, B.K.; Kim, Y.; Cho, T. Low-frequency impact sound transmission of floating floor: Case study of mortar bed on concrete slab with continuous interlayer. *Build. Environ.* **2015**, *94*, 793–801. [CrossRef]
26. Feng, L. Modified impedance tube measurements and energy dissipation inside absorptive materials. *Appl. Acoust.* **2013**, *74*, 1480–1485. [CrossRef]
27. ASTM. *E413 Classification for Rating Sound Insulation*; ASTM International: West Conshohocken, PA, USA, 2011.
28. Pérez, M.; Fuente, M. Acoustic design through predictive methods in Cross Laminated Timber (CLT) panel structures for buildings. In Proceedings of the 42nd International Congress and Exposition on Noise Control Engineering 2013, Innsbruck, Austria, 15–18 September 2013. INTER-NOISE 2013: Noise Control for Quality of Life.
29. Çelik, O.N.; Atiş, C.D. Compactibility of hot bituminous mixtures made with crumb rubber-modified binders. *Constr. Build. Mater.* **2008**, *22*, 1143–1147. [CrossRef]
30. Ma, Y.; Wang, H.; Li, H.; Hong, J. Study on metal rubber material's characteristics of damping and sound absorption. In Proceedings of the ASME Turbo Expo, Berlin, Germany, 9–13 June 2008.
31. Ghizdăveț, Z.; Ștefan, B.M.; Nastac, D.; Vasile, O.; Bratu, M. Sound absorbing materials made by embedding crumb rubber waste in a concrete matrix. *Constr. Build. Mater.* **2016**, *124*, 755–763. [CrossRef]
32. Asdrubali, F.; D'Alessandro, F.; Schiavoni, S. Sound absorbing properties of materials made of rubber crumbs. *J. Acoust. Soc. Am.* **2008**, *123*, 3037. [CrossRef]
33. Cox, T.J.; D'Antonio, P. *Acoustic Absorbers and Diffusers, Theory, Design and Application, Third Edition*; CRC Press: Boca Raton, FL, USA, 2017; ISBN 9781498740999.
34. Ibrahim, R.A. Recent advances in nonlinear passive vibration isolators. *J. Sound Vib.* **2008**, *314*, 371–452. [CrossRef]

35. Sun, Z.; Shen, Z.; Ma, S.; Zhang, X. Sound absorption application of fiberglass recycled from waste printed circuit boards. *Mater. Struct. Constr.* **2015**, *48*, 387–392. [CrossRef]
36. Öqvist, R.; Ljunggren, F.; Johnsson, R. Walking sound annoyance vs. impact sound insulation from 20 Hz. *Appl. Acoust.* **2018**, *135*, 1–7. [CrossRef]
37. Fediuk, R.; Yushin, A. Composite binders for concrete with reduced permeability. In Proceedings of the IOP Conference Series: Materials Science and Engineering, Tomsk, Russia, 9–11 November 2015; IOP Publishing: Bristol, UK, 2016; Volume 116.
38. Neithalath, N.; Weiss, J.; Olek, J. Characterizing Enhanced Porosity Concrete using electrical impedance to predict acoustic and hydraulic performance. *Cem. Concr. Res.* **2006**, *36*, 2074–2085. [CrossRef]
39. Asdrubali, F.; Schiavoni, S.; Horoshenkov, K.V. A review of sustainable materials for acoustic applications. *Build. Acoust.* **2012**, *19*, 283–311. [CrossRef]
40. Kim, H.K.; Lee, H.K. Influence of cement flow and aggregate type on the mechanical and acoustic characteristics of porous concrete. *Appl. Acoust.* **2010**, *71*, 607–615. [CrossRef]
41. Arenas, C.; Leiva, C.; Vilches, L.F.; Cifuentes, H.; Rodríguez-Galán, M. Technical specifications for highway noise barriers made of coal bottom ash-based sound absorbing concrete. *Constr. Build. Mater.* **2015**, *95*, 585–591. [CrossRef]
42. ASTM. *C423 Standard Test Method for Sound Absorption and Sound Absorption Coefficients by the Reverberation Room Method 1*; ASTM International: West Conshohocken, PA, USA, 2007.
43. Version, D.; Materials, F.; Materials, S.; Materials, W. Absorption Coefficients α of Building Materials and Finishes More Tables and Charts of Absorption Coefficients. 2020. Available online: http://www.sengpielaudio.com/calculator-RT60Coeff.htm (accessed on 21 October 2020).
44. Tiwari, V.; Shukla, A.; Bose, A. Acoustic properties of cenosphere reinforced cement and asphalt concrete. *Appl. Acoust.* **2004**, *65*, 263–275. [CrossRef]
45. Sukontasukkul, P. Use of crumb rubber to improve thermal and sound properties of pre-cast concrete panel. *Constr. Build. Mater.* **2009**, *23*, 1084–1092. [CrossRef]
46. Ling, T.C.; Nor, H.M.; Lim, S.K. Using recycled waste tyres in concrete paving blocks. In *Proceedings of the Institution of Civil Engineers-Waste and Resource Management*; Thomas Telford Ltd.: London, UK, 2010.
47. Glé, P.; Gourdon, E.; Arnaud, L. Acoustical properties of materials made of vegetable particles with several scales of porosity. *Appl. Acoust.* **2011**, *72*, 249–259. [CrossRef]
48. Leiva, C.; Vilches, L.F.; Arenas, C.; Delgado, S.; Fernández-Pereira, C. Potential recycling of bottom and fly ashes in acoustic mortars and concretes. *ACI Mater. J.* **2012**, *109*, 529.
49. Arenas, C.; Leiva, C.; Vilches, L.F.; Cifuentes, H. Use of co-combustion bottom ash to design an acoustic absorbing material for highway noise barriers. *Waste Manag.* **2013**, *33*, 2316–2321. [CrossRef]
50. Buratti, C.; Moretti, E.; Belloni, E.; Agosti, F. Development of innovative aerogel based plasters: Preliminary thermal and acoustic performance evaluation. *Sustainability* **2014**, *6*, 5839–5852. [CrossRef]
51. Zhang, Z.; Provis, J.L.; Reid, A.; Wang, H. Mechanical, thermal insulation, thermal resistance and acoustic absorption properties of geopolymer foam concrete. *Cem. Concr. Compos.* **2015**, *62*, 97–105. [CrossRef]
52. Carbajo, J.; Esquerdo-Lloret, T.V.; Ramis, J.; Nadal-Gisbert, A.V.; Denia, F.D. Acoustic properties of porous concrete made from arlite and vermiculite lightweight aggregates. *Mater. Constr.* **2015**, *65*, 72. [CrossRef]
53. Kinnane, O.; Reilly, A.; Grimes, J.; Pavia, S.; Walker, R. Acoustic absorption of hemp-lime construction. *Constr. Build. Mater.* **2016**, *122*, 674–682. [CrossRef]
54. Flores Medina, N.; Flores-Medina, D.; Hernández-Olivares, F. Influence of fibers partially coated with rubber from tire recycling as aggregate on the acoustical properties of rubberized concrete. *Constr. Build. Mater.* **2016**, *129*, 25–36. [CrossRef]
55. Chen, Y.; Yu, Q.L.; Brouwers, H.J.H. Acoustic performance and microstructural analysis of bio-based lightweight concrete containing miscanthus. *Constr. Build. Mater.* **2017**, *157*, 839–851. [CrossRef]
56. Gourlay, E.; Glé, P.; Marceau, S.; Foy, C.; Moscardelli, S. Effect of water content on the acoustical and thermal properties of hemp concretes. *Constr. Build. Mater.* **2017**, *139*, 513–523. [CrossRef]
57. Carbajo, J.; Esquerdo-Lloret, T.V.; Ramis, J.; Nadal-Gisbert, A.V.; Denia, F.D. Acoustic modeling of perforated concrete using the dual porosity theory. *Appl. Acoust.* **2017**, *115*, 150–157. [CrossRef]
58. Stolz, J.; Boluk, Y.; Bindiganavile, V. Mechanical, thermal and acoustic properties of cellular alkali activated fly ash concrete. *Cem. Concr. Compos.* **2018**, *94*, 24–32. [CrossRef]
59. Mastali, M.; Kinnunen, P.; Isomoisio, H.; Karhu, M.; Illikainen, M. Mechanical and acoustic properties of fiber-reinforced alkali-activated slag foam concretes containing lightweight structural aggregates. *Constr. Build. Mater.* **2018**, *187*, 371–381. [CrossRef]
60. Oancea, I.; Bujoreanu, C.; Budescu, M.; Benchea, M.; Grădinaru, C.M. Considerations on sound absorption coefficient of sustainable concrete with different waste replacements. *J. Clean. Prod.* **2018**, *203*, 301–312. [CrossRef]
61. Fernea, R.; Manea, D.L.; Plesa, L.; Iernutan, R.; Dumitran, M. Acoustic and thermal properties of hemp-cement building materials. *Procedia Manuf.* **2019**, *32*, 208–215. [CrossRef]
62. Mhaya, A.M.; Huseien, G.F.; Abidin, A.R.Z.; Ismail, M. Long-term mechanical and durable properties of waste tires rubber crumbs replaced GBFS modified concretes. *Constr. Build. Mater.* **2020**, *256*, 119505. [CrossRef]

63. Neves, E.; Sousa, A.; Gibbs, B.M. Low frequency impact sound transmission in dwellings through homogeneous concrete floors and floating floors. *Appl. Acoust.* **2011**, *72*, 177–189. [CrossRef]
64. Yang, Z.; Dai, H.M.; Chan, N.H.; Ma, G.C.; Sheng, P. Acoustic metamaterial panels for sound attenuation in the 50–1000 Hz regime. *Appl. Phys. Lett.* **2010**, *96*, 041906. [CrossRef]
65. Tolstoy, A.D.; Lesovik, V.S.; Glagolev, E.S.; Krymova, A.I. Synergetics of hardening construction systems. In *IOP Conference Series: Materials Science and Engineering*; IOP Publishing: Bristol, UK, 2018.
66. Azkorra, Z.; Pérez, G.; Coma, J.; Cabeza, L.F.; Bures, S.; Álvaro, J.E.; Erkoreka, A.; Urrestarazu, M. Evaluation of green walls as a passive acoustic insulation system for buildings. *Appl. Acoust.* **2015**, *89*, 46–56. [CrossRef]
67. Akishita, S.; Mitani, A.; Uchida, H.; Shioda, M. Acoustically evaluating measurement of active insulation panel of floor impact noise. In Proceedings of the International Congress on Noise Control Engineering, Rio de Janeiro, Brazil, 7–10 August 2005.
68. Maa, D.-Y. Potential of microperforated panel absorber. *J. Acoust. Soc. Am.* **1998**, *104*, 2861–2866. [CrossRef]
69. Park, S.W.; Lee, I.H.; Ko, K.I.; Lee, E.T. Evaluation of the pull-out strength of connections with roof cladding using honeycomb panels with cool roof performance capacity. *Int. J. Steel Struct.* **2016**, *16*, 505–516. [CrossRef]
70. Vatin, N.I.; Pestryakov, I.I.; Kiski, S.S.; Teplova, Z.S. Influence of the geometrical values of hollowness on the physicotechnical characteristics of the concrete vibropressed wall stones. In *Applied Mechanics and Materials*; Trans Tech Publications Ltd.: Stafa-Zurich, Switzerland, 2014.
71. Ayzenshtadt, A.; Lesovik, V.; Frolova, M.; Tutygin, A.; Danilov, V. Nanostructured wood mineral composite. *Procedia Eng.* **2015**, *117*, 45–51. [CrossRef]
72. Narayanan, N.; Ramamurthy, K. Structure and properties of aerated concrete: A review. *Cem. Concr. Compos.* **2000**, *22*, 321–329. [CrossRef]
73. Kwon, H.B.; Lee, C.W.; Jun, B.S.; Yun, J.D.; Weon, S.Y.; Koopman, B. Recycling waste oyster shells for eutrophication control. *Resour. Conserv. Recycl.* **2004**, *41*, 75–82. [CrossRef]
74. Laukaitis, A.; Fiks, B. Acoustical properties of aerated autoclaved concrete. *Appl. Acoust.* **2006**, *67*, 284–296. [CrossRef]
75. Luna-Galiano, Y.; Leiva, C.; Arenas, C.; Fernández-Pereira, C. Fly ash based geopolymeric foams using silica fume as pore generation agent. Physical, mechanical and acoustic properties. *J. Non. Cryst. Solids* **2018**, *500*, 196–204. [CrossRef]
76. ISO. *ISO 10534-2:1998—Acoustics—Determination of Sound Absorption Coefficient and Impedance in Impedance Tubes—Part 2: Transfer-Function Method*; ISO: Geneva, Switzerland, 1998.
77. Wu, J.; Zhang, Z.; Zhang, Y.; Li, D. Preparation and characterization of ultra-lightweight foamed geopolymer (UFG) based on fly ash-metakaolin blends. *Constr. Build. Mater.* **2018**, *168*, 771–779. [CrossRef]
78. Jones, M.R.; Ozlutas, K.; Ouzounidou, A.; Rathbone, R.F. Behaviour of PC-CSA-FA Blends in Foamed Concrete. In Proceedings of the Conference World Coal Ash Conference, Lexington, MA, USA, 22–25 April 2013.
79. Park, S.B.; Seo, D.S.; Lee, J. Studies on the sound absorption characteristics of porous concrete based on the content of recycled aggregate and target void ratio. *Cem. Concr. Res.* **2005**, *35*, 1846–1854. [CrossRef]
80. Erofeev, V.T.; Makridin, N.I.; Maksimova, I.N. Mechanical and acoustic behavior of modified structures of a cement stone of different age. *Izv. Vyss. Uchebnykh Zaved. Seriya Teknol. Tekst. Promyshlennosti* **2019**, *4*, 74–79.
81. Provis, J.L.; Myers, R.J.; White, C.E.; Rose, V.; Van Deventer, J.S.J. X-ray microtomography shows pore structure and tortuosity in alkali-activated binders. *Cem. Concr. Res.* **2012**, *42*, 855–864. [CrossRef]
82. Blanco, F.; Garcı'a, P.; Mateos, P.; Ayala, J. Characteristics and properties of lightweight concrete manufactured with cenospheres. *Cem. Concr. Res.* **2000**, *30*, 1715–1722. [CrossRef]
83. Luo, X.; Li, W.; Jin, X.; Zeng, L. Effects of porosity and pore size on sound absorption characteristic of ceramsite porous material. *J. Chinese Ceram. Soc.* **2011**, *39*, 158–163.
84. Amran, M.; Debbarma, S.; Ozbakkaloglu, T. Fly ash-based eco-friendly geopolymer concrete: A critical review of the long-term durability properties. *Constr. Build. Mater.* **2021**, *270*, 121857. [CrossRef]
85. Guo, Y.C.; Zhang, J.H.; Chen, G.M.; Xie, Z.H. Compressive behaviour of concrete structures incorporating recycled concrete aggregates, rubber crumb and reinforced with steel fibre, subjected to elevated temperatures. *J. Clean. Prod.* **2014**, *72*, 193–203. [CrossRef]
86. Lesovik, V.S.; Pershina, I.L. Acoustic Factor in the Formation of Architectural Space. In Proceedings of the IOP Conference Series: Materials Science and Engineering, Zawiercie, Poland, 26–29 September 2018.
87. Paje, S.E.; Bueno, M.; Terán, F.; Miró, R.; Pérez-Jiménez, F.; Martínez, A.H. Acoustic field evaluation of asphalt mixtures with crumb rubber. *Appl. Acoust.* **2010**, *71*, 578–582. [CrossRef]
88. Lesovik, V.S.; Bessonov, I.V.; Bulgakov, B.I.; Larsen, O.A.; Puchka, O.V.; Vaysera, S.S. Approach on Improving the Performance of Thermal Insulating and Acoustic Glass Composites. In Proceedings of the IOP Conference Series: Materials Science and Engineering, Zawiercie, Poland, 26–29 September 2018.
89. Factors Influencing Acoustic Performance of Sound Absorptive Materials. *Aust. J. Basic Appl. Sci.* **2009**, *3*, 4610–4617.
90. Wang, X.; Chin, C.S.; Xia, J. Material characterization for sustainable concrete paving blocks. *Appl. Sci.* **2019**, *9*, 1197. [CrossRef]
91. Hamoush, S.; Abu-Lebdeh, T.; Picornell, M.; Amer, S. Development of sustainable engineered stone cladding for toughness, durability, and energy conservation. *Constr. Build. Mater.* **2011**, *25*, 4006–4016. [CrossRef]
92. Najim, K.B.; Hall, M.R. Crumb rubber aggregate coatings/pre-treatments and their effects on interfacial bonding, air entrapment and fracture toughness in self-compacting rubberised concrete (SCRC). *Mater. Struct. Constr.* **2013**, *46*, 2029–2043. [CrossRef]

93. Richardson, A.E.; Coventry, K.A.; Ward, G. Freeze/thaw protection of concrete with optimum rubber crumb content. *J. Clean. Prod.* **2012**, *23*, 96–103. [CrossRef]
94. Lesovik, V.S.; Elistratkin, M.Y.; Glagolev, E.S.; Voronov, V.V.; Absimetov, M.V. Non-autoclaved aerated concrete on the basis of composite binder using technogenic raw materials. *Mater. Sci. Forum* **2018**, *945*, 205–211. [CrossRef]
95. Wakefield Acoustics Ltd. *City of Vancouver Noise Control Manual City of Vancouver Noise Control Manual*; Wakefield Acoustics Ltd., Engineering Services: Vancouver, BC, Canada, 2005.
96. Rindel, J.H. Acoustic Quality and Sound Insulation between Dwellings. *Build. Acoust.* **1998**, *5*, 291–301. [CrossRef]
97. Chen, B.; Liu, J.; Chen, L.Z. Experimental study of lightweight expanded polystyrene aggregate concrete containing silica fume and polypropylene fibers. *J. Shanghai Jiaotong Univ.* **2010**, *15*, 129–137. [CrossRef]
98. Sikora, J.; Turkiewicz, J. Sound absorption coefficients of granular materials. *Mech. Control* **2010**, *29*, 149–157.
99. Olukunle, B.G.; Uche, N.B.; Efomo, A.O.; Adeyemi, G.A.; Joshua, J.K. Data on acoustic behaviour of coconut fibre-reinforced concrete. *Data Br.* **2018**, *21*, 1004–1007. [CrossRef]
100. Zulkifli, R.; Mohd Nor, M.J.; Mat Tahir, M.F.; Ismail, A.R.; Nuawi, M.Z. Acoustic properties of multi-layer coir fibres sound absorption panel. *J. Appl. Sci.* **2008**, *8*, 3709–3714. [CrossRef]
101. Lesovik, V.; Voronov, V.; Glagolev, E.; Fediuk, R.; Alaskhanov, A.; Amran, M.; Murali, G.; Baranov, A. Improving the behaviors of foam concrete through the use of composite binder. *J. Build. Eng.* **2020**, *13*, 101414. [CrossRef]
102. Klyuev, S.V.; Klyuev, A.V.; Khezhev, T.A.; Pukharenko, Y.V. High-strength fine-grained fiber concrete with combined reinforcement by fiber. *J. Eng. Appl. Sci.* **2018**, *13*, 6407–6412.
103. Kim, H.K.; Jeon, J.H.; Lee, H.K. Workability, and mechanical, acoustic and thermal properties of lightweight aggregate concrete with a high volume of entrained air. *Constr. Build. Mater.* **2012**, *29*, 193–200. [CrossRef]
104. Navacerrada, M.A.; Díaz, C.; Pedrero, A.; García, L.E. Acoustic properties of aluminium foams. *Mater. Constr.* **2008**, *58*, 85–98.
105. Arenas, C.; Leiva, C.; Vilches, L.F.; González Ganso, J.A. Approaching a methodology for the development of a multilayer sound absorbing device recycling coal bottom ash. *Appl. Acoust.* **2017**, *115*, 81–87. [CrossRef]
106. Parker, G. Effective Noise Barrier Design and Specification Chair CEN TC226/WG6/TG1 for Road Traffic Noise reducing Devices. In Proceedings of the Acoustics, Christchurch, New Zealand, 20–22 November 2006.
107. Lesovik, V.; Volodchenko AFediuk, R.; Aman, M.; Timokhin, R. Enhancing performances of clay masonry materials based on nanosize mine waste. *Constr. Build. Mater.* **2021**, *269*, 121333. [CrossRef]
108. Wang, C.N.; Torng, J.H. Experimental study of the absorption characteristics of some porous fibrous materials. *Appl. Acoust.* **2001**, *62*, 447–459. [CrossRef]
109. Chen, Y.; Jiang, N. Carbonized and Activated Non-wovens as High-Performance Acoustic Materials: Part I Noise Absorption. *Text. Res. J.* **2007**, *77*, 785–791. [CrossRef]
110. Sun, F.; Chen, H.; Wu, J.; Feng, K. Sound absorbing characteristics of fibrous metal materials at high temperatures. *Appl. Acoust.* **2010**, *71*, 221–235. [CrossRef]
111. Kino, N. A comparison of two acoustical methods for estimating parameters of glass fibre and melamine foam materials. *Appl. Acoust.* **2012**, *73*, 590–603. [CrossRef]
112. Lippitz, N.; Rösler, J.; Hinze, B. Potential of metal fibre felts as passive absorbers in absorption silencers. *Metals* **2013**, *3*, 150–158. [CrossRef]
113. Prabhakaran, S.; Krishnaraj, V.; Senthil Kumar, M.; Zitoune, R. Sound and vibration damping properties of flax fiber reinforced composites. *Procedia Eng.* **2014**, *97*, 573–581. [CrossRef]
114. Wang, X.; Li, Y.; Chen, T.; Ying, Z. Research on the sound absorption characteristics of porous metal materials at high sound pressure levels. *Adv. Mech. Eng.* **2015**, *7*, 1687814015575429. [CrossRef]
115. Yang, Y.; Chen, Z.; Chen, Z.; Fu, R.; Li, Y. Sound insulation properties of sandwich structures on glass fiber felts. *Fibers Polym.* **2015**, *16*, 1568–1577. [CrossRef]
116. Kim, B.S.; Cho, S.J.; Min, D.K.; Park, J. Sound absorption structure in helical shapes made using fibrous paper. *Compos. Struct.* **2015**, *134*, 90–94. [CrossRef]
117. Lippitz, N.; Blech, C.; Langer, S.; Rösler, J. Identification of material parameters for the simulation of acoustic absorption of fouled sintered fiber felts. *Materials* **2016**, *9*, 709. [CrossRef]
118. Moretti, E.; Belloni, E.; Agosti, F. Innovative mineral fiber insulation panels for buildings: Thermal and acoustic characterization. *Appl. Energy* **2016**, *169*, 421–432. [CrossRef]
119. Wang, R.; Yan, R.; Lou, C.W.; Lin, J.H. Characterization of acoustic-absorbing inter/intra-ply hybrid laminated composites under dynamic loading. *Fibers Polym.* **2016**, *17*, 439–452. [CrossRef]
120. Yang, Y.; Li, B.; Chen, Z.; Sui, N.; Chen, Z.; Saeed, M.U.; Li, Y.; Fu, R.; Wu, C.; Jing, Y. Acoustic properties of glass fiber assembly-filled honeycomb sandwich panels. *Compos. Part B Eng.* **2016**, *96*, 281–286. [CrossRef]
121. Zhu, J.; Sun, J.; Tang, H.; Wang, J.; Ao, Q.; Bao, T.; Song, W. Gradient-structural optimization of metal fiber porous materials for sound absorption. *Powder Technol.* **2016**, *301*, 1235–1241. [CrossRef]
122. ISO. *ISO 354. International Organization for Standardization Measurement of Sound Absorption in a Reverberation Room*; ISO: Geneva, Switzerland, 2003.
123. Peceño, B.; Arenas, C.; Alonso-Fariñas, B.; Leiva, C. Substitution of Coarse Aggregates with Mollusk-Shell Waste in Acoustic-Absorbing Concrete. *J. Mater. Civ. Eng.* **2019**, *31*, 04019077. [CrossRef]

124. Paridah, M.T.; Juliana, A.H.; El-Shekeil, Y.A.; Jawaid, M.; Alothman, O.Y. Measurement of mechanical and physical properties of particleboard by hybridization of kenaf with rubberwood particles. *Meas. J. Int. Meas. Confed.* **2014**, *56*, 70–80. [CrossRef]
125. Zhong, R.; Xu, M.; Vieira Netto, R.; Wille, K. Influence of pore tortuosity on hydraulic conductivity of pervious concrete: Characterization and modeling. *Constr. Build. Mater.* **2016**, *125*, 1158–1168. [CrossRef]
126. König, J.; Lopez-Gil, A.; Cimavilla-Roman, P.; Rodriguez-Perez, M.A.; Petersen, R.R.; Østergaard, M.B.; Iversen, N.; Yue, Y.; Spreitzer, M. Synthesis and properties of open- and closed-porous foamed glass with a low density. *Constr. Build. Mater.* **2020**, *247*, 118574. [CrossRef]
127. Kyaw Oo D'Amore, G.; Caniato, M.; Travan, A.; Turco, G.; Marsich, L.; Ferluga, A.; Schmid, C. Innovative thermal and acoustic insulation foam from recycled waste glass powder. *J. Clean. Prod.* **2017**, *165*, 1306–1315. [CrossRef]
128. Mohajerani, A.; Vajna, J.; Cheung, T.H.H.; Kurmus, H.; Arulrajah, A.; Horpibulsuk, S. Practical recycling applications of crushed waste glass in construction materials: A review. *Constr. Build. Mater.* **2017**, *156*, 443–467. [CrossRef]
129. Meyer, C. The greening of the concrete industry. *Cem. Concr. Compos.* **2009**, *31*, 601–605. [CrossRef]
130. Kamseu, E.; Ponzoni, C.; Tippayasam, C.; Taurino, R.; Chaysuwan, D.; Sglavo, V.M.; Thavorniti, P.; Leonelli, C. Self-compacting geopolymer concretes: Effects of addition of aluminosilicate-rich fines. *J. Build. Eng.* **2016**, *5*, 211–221. [CrossRef]
131. Li, J.; Zhuang, X.; Monfort, E.; Querol, X.; Llaudis, A.S.; Font, O.; Moreno, N.; Ten, F.J.G.; Izquierdo, M. Utilization of coal fly ash from a Chinese power plant for manufacturing highly insulating foam glass: Implications of physical, mechanical properties and environmental features. *Constr. Build. Mater.* **2018**, *175*, 64–76. [CrossRef]
132. Klyuev, S.V.; Klyuev, A.V.; Vatin, N.I. Fiber concrete for the construction industry. *Mag. Civ. Eng.* **2018**, *84*, 41–47.
133. Yoo, D.Y.; Banthia, N.; Yoon, Y.S. Flexural behavior of ultra-high-performance fiber-reinforced concrete beams reinforced with GFRP and steel rebars. *Eng. Struct.* **2016**, *111*, 246–262. [CrossRef]
134. Kongsong, P.; Sikong, L.; Niyomwas, S.; Rachpech, V. Photocatalytic antibacterial performance of glass fibers thin film coated with N-doped SnO2/TiO2. *Sci. World, J.* **2014**, *1*, 869706.
135. Joshi, S.V.; Drzal, L.T.; Mohanty, A.K.; Arora, S. Are natural fiber composites environmentally superior to glass fiber reinforced composites? *Compos. Part A Appl. Sci. Manuf.* **2004**, *35*, 371–376. [CrossRef]
136. Hesky, D.; Aneziris, C.G.; Groß, U.; Horn, A. Water and waterglass mixtures for foam glass production. *Ceram. Int.* **2015**, *41*, 12604–12613. [CrossRef]
137. Karandashova, N.S.; Goltsman, B.M.; Yatsenko, E.A. Analysis of Influence of Foaming Mixture Components on Structure and Properties of Foam Glass. *IOP Conf. Ser. Mater. Sci. Eng.* **2017**, *262*, 012020. [CrossRef]
138. Qu, Y.N.; Xu, J.; Su, Z.G.; Ma, N.; Zhang, X.Y.; Xi, X.Q.; Yang, J.L. Lightweight and high-strength glass foams prepared by a novel green spheres hollowing technique. *Ceram. Int.* **2016**, *42*, 2370–2377. [CrossRef]
139. Petersen, R.R.; König, J.; Yue, Y. The mechanism of foaming and thermal conductivity of glasses foamed with MnO_2. *J. Non. Cryst. Solids* **2015**, *425*, 74–82. [CrossRef]
140. König, J.; Petersen, R.R.; Yue, Y. Fabrication of highly insulating foam glass made from CRT panel glass. *Ceram. Int.* **2015**, *41*, 9793–9800. [CrossRef]
141. König, J.; Petersen, R.R.; Yue, Y. Influence of the glass particle size on the foaming process and physical characteristics of foam glasses. *J. Non. Cryst. Solids* **2016**, *447*, 190–197. [CrossRef]
142. Fernandes, H.R.; Andreola, F.; Barbieri, L.; Lancellotti, I.; Pascual, M.J.; Ferreira, J.M.F. The use of egg shells to produce Cathode Ray Tube (CRT) glass foams. *Ceram. Int.* **2013**, *39*, 9071–9078. [CrossRef]
143. Huo, W.; Yan, S.; Wu, J.M.; Liu, J.; Chen, Y.; Qu, Y.; Tang, X.; Yang, J. A novel fabrication method for glass foams with small pore size and controllable pore structure. *J. Am. Ceram. Soc.* **2017**, *100*, 5502–5511. [CrossRef]
144. König, J.; Nemanič, V.; Žumer, M.; Petersen, R.R.; Østergaard, M.B.; Yue, Y.; Suvorov, D. Evaluation of the contributions to the effective thermal conductivity of an open-porous-type foamed glass. *Constr. Build. Mater.* **2019**, *214*, 337–343. [CrossRef]
145. Xu, Q.; Zeng, J.; Li, X.; Xu, J.; Liu, X. 3D nano-macroporous structured TiO2-foam glass as an efficient photocatalyst for organic pollutant treatment. *RSC Adv.* **2016**, *6*, 51888–51893. [CrossRef]
146. Rincón, A.; Giacomello, G.; Pasetto, M.; Bernardo, E. Novel 'inorganic gel casting' process for the manufacturing of glass foams. *J. Eur. Ceram. Soc.* **2017**, *37*, 2227–2234. [CrossRef]
147. König, J.; Petersen, R.R.; Yue, Y. Influence of the glass-calcium carbonate mixture's characteristics on the foaming process and the properties of the foam glass. *J. Eur. Ceram. Soc.* **2014**, *34*, 1591–1598. [CrossRef]
148. Østergaard, M.B.; Petersen, R.R.; König, J.; Bockowski, M.; Yue, Y. Foam glass obtained through high-pressure sintering. *J. Am. Ceram. Soc.* **2018**, *101*, 3917–3923. [CrossRef]
149. Wang, B.; Matsumaru, K.; Yang, J.; Fu, Z.; Ishizaki, K. Mechanical behavior of cellular borosilicate glass with pressurized Ar-filled closed pores. *Acta Mater.* **2012**, *60*, 4185–4193. [CrossRef]
150. König, J.; Petersen, R.R.; Iversen, N.; Yue, Y. Suppressing the effect of cullet composition on the formation and properties of foamed glass. *Ceram. Int.* **2018**, *44*, 11143–11150. [CrossRef]
151. Østergaard, M.B.; Petersen, R.R.; König, J.; Yue, Y. Effect of alkali phosphate content on foaming of CRT panel glass using Mn3O4 and carbon as foaming agents. *J. Non. Cryst. Solids* **2018**, *482*, 217–222. [CrossRef]
152. Østergaard, M.B.; Cai, B.; Petersen, R.R.; König, J.; Lee, P.D.; Yue, Y. Impact of pore structure on the thermal conductivity of glass foams. *Mater. Lett.* **2019**, *250*, 72–74. [CrossRef]
153. Caniato, M.; Bettarello, F.; Taffarel, M. Sound power level of speaking people. *Acoustical Society of America* **2013**, *19*, 040026.

154. Caniato, M.; Bettarello, F.; Fausti, P.; Marsich, L.; Ferluga, A.; Schmid, C. Low frequency noise and disturbance assessment methods: A brief literature overview and a new proposal. *Acoustical Society of America* **2016**, *28*, 032001.
155. Caniato, M.; Bettarello, F.; Schmid, C.; Fausti, P. Assessment criterion for indoor noise disturbance in the presence of low frequency sources. *Appl. Acoust.* **2016**, *113*, 22–33. [CrossRef]
156. Bonfiglio, P.; Pompoli, F. Frequency-dependent mechanical modelling of poro-elastic materials for sound transmission loss simulations. *Noise Control Eng. J.* **2016**, *64*, 627–633. [CrossRef]
157. Bonfiglio, P.; Pompoli, F. Frequency dependent tortuosity measurement by means of ultrasonic tests. In Proceedings of the 14th International Congress on Sound and Vibration 2007, ICSV 2007, Cairns, Australia, 9–12 July 2007.
158. Di Bella, A.; Granzotto, N.; Pavarin, C. Comparative analysis of thermal and acoustic performance of building elements. In Proceedings of the Forum Acusticum, Krakow, Poland, 7–12 September 2014.
159. Asdrubali, F.; D'Alessandro, F.; Schiavoni, S. A review of unconventional sustainable building insulation materials. *Sustain. Mater. Technol.* **2015**, *4*, 1–17. [CrossRef]
160. Mihai, T.; Iordache, V. Determining the Indoor Environment Quality for an Educational Building. *Energy Procedia* **2016**, *85*, 566–574. [CrossRef]
161. Schiavoni, S.; D'Alessandro, F.; Bianchi, F.; Asdrubali, F. Insulation materials for the building sector: A review and comparative analysis. *Renew. Sustain. Energy Rev.* **2016**, *62*, 988–1011. [CrossRef]
162. Roulet, C.A.; Johner, N.; Foradini, F.; Bluyssen, P.; Cox, C.; De Oliveira Fernandes, E.; Müller, B.; Aizlewood, C. Perceived health and comfort in relation to energy use and building characteristics. *Build. Res. Inf.* **2006**, *34*, 467–474. [CrossRef]
163. Huang, Y.; Duan, X.; Cui, Y.; Lauhon, L.J.; Kim, K.H.; Lieber, C.M. Logic gates and computation from assembled nanowire building blocks. *Science* **2001**, *294*, 1313–1317. [CrossRef]
164. Ismail, M.R. Quiet environment: Acoustics of vertical green wall systems of the Islamic urban form. *Front. Archit. Res.* **2013**, *2*, 162–177. [CrossRef]
165. Rasmussen, B. International proposal for an acoustic classification scheme for dwellings—Background and perspectives. In Proceedings of the INTERNOISE 2014—43rd International Congress on Noise Control Engineering: Improving the World Through Noise Control, Melbourne, Australia, 16–19 November 2014.
166. Calleri, C.; Astolfi, A.; Shtrepi, L.; Prato, A.; Schiavi, A.; Zampini, D.; Volpatti, G. Characterization of the sound insulation properties of a two-layers lightweight concrete innovative façade. *Appl. Acoust.* **2019**, *145*, 267–277. [CrossRef]
167. Isolamento, R.E.; Di, P. Modulo a getto singolo. *ECOSISM Adv. Build. Technol.* **2020**, *5*, 267–277.
168. Hongisto, V.; Mäkilä, M.; Suokas, M. Satisfaction with sound insulation in residential dwellings—The effect of wall construction. *Build. Environ.* **2015**, *85*, 309–320. [CrossRef]
169. Wittstock, V. Determination of measurement uncertainties in building acoustics by interlaboratory tests. Part 1: Airborne sound insulation. *Acta Acust. United Acust.* **2015**, *101*, 88–98. [CrossRef]
170. Martinović, A.; Mathias, M.; Weissenberg, J.; Van Gool, L. A three-layered approach to facade parsing. In Proceedings of the European Conference on Computer Vision, Florence, Italy, 7–13 October 2012.
171. Müller, P.; Zeng, G.; Wonka, P.; Van Gool, L. Image-based procedural modeling of facades. *ACM Trans. Graph.* **2007**, *26*, 85. [CrossRef]
172. Ljunggren, F.; Simmons, C.; Hagberg, K. Findings from the AkuLite project: Correlation between measured vibro—Acoustic parameters and subjective perception in lightweight buildings. In Proceedings of the International Congress and Exposition on Noise Control Engineering, Innsbruck, Austria, 15–18 September 2013.
173. Moussavi Nadoushani, Z.S.; Akbarnezhad, A.; Ferre Jornet, J.; Xiao, J. Multi-criteria selection of façade systems based on sustainability criteria. *Build. Environ.* **2017**, *121*, 67–78. [CrossRef]
174. Orion-SD Porcelain Stoneware Ventilated Facade. Vent. *Facades 2020.* Available online: https://orion-sd.ru/raboty/vent-fasad-2 (accessed on 22 August 2020).
175. Frazão, C.; Barros, J.; Toledo Filho, R.; Ferreira, S.; Gonçalves, D. Development of sandwich panels combining Sisal Fiber-Cement Composites and Fiber-Reinforced Lightweight Concrete. *Cem. Concr. Compos.* **2018**, *86*, 206–223. [CrossRef]
176. Cuypers, H.; Wastiels, J. Analysis and verification of the performance of sandwich panels with textile reinforced concrete faces. *J. Sandw. Struct. Mater.* **2011**, *13*, 589–603. [CrossRef]
177. Flores-Johnson, E.A.; Li, Q.M. Structural behaviour of composite sandwich panels with plain and fibre-reinforced foamed concrete cores and corrugated steel faces. *Compos. Struct.* **2012**, *94*, 1555–1563. [CrossRef]
178. Jeon, J.Y.; Hong, J.Y.; Kim, S.M.; Lee, P.J. Classification of heavy-weight floor impact sounds in multi-dwelling houses using an equal-appearing interval scale. *Build. Environ.* **2015**, *94*, 821–828. [CrossRef]
179. Zhang, B.; Poon, C.S. Sound insulation properties of rubberized lightweight aggregate concrete. *J. Clean. Prod.* **2018**, *172*, 3176–3185. [CrossRef]
180. Lee, S.; Hwang, D.; Park, J.; Jeon, J.Y. Cause and perception of amplitude modulation of heavy-weight impact sounds in concrete wall structures. *Build. Environ.* **2015**, *94*, 785–792. [CrossRef]
181. Abirami, T.; Loganaganandan, M.; Murali, G.; Fediuk, R.; Vickhram Sreekrishna, R.; Vignesh, T.; Januppriya, G.; Karthikeyan, K. Experimental research on impact response of novel steel fibrous concretes under falling mass impact. *Constr. Build. Mater.* **2019**, *222*, 447–457. [CrossRef]

182. Di, J.; Sun, Y.; Yu, K.; Liu, L.; Qin, F. Experimental investigation of shear performance of existing PC hollow slab. *Eng. Struct.* **2020**, *211*, 110451. [CrossRef]
183. Nguyen, H.T.N.; Tan, K.H.; Kanda, T. Experimental and numerical studies on shear behavior of deep prestressed concrete hollow core slabs. In Proceedings of the High Tech Concrete: Where Technology and Engineering Meet, Maastricht, The Netherlands, 12–14 June 2017.
184. Al-Rubaye, M.; Manalo, A.; Alajarmeh, O.; Ferdous, W.; Lokuge, W.; Benmokrane, B.; Edoo, A. Flexural behaviour of concrete slabs reinforced with GFRP bars and hollow composite reinforcing systems. *Compos. Struct.* **2020**, *236*, 111836. [CrossRef]
185. Rivera, U. Precast Decking System. *Cobute Concr. Build. Tech.* 2020. Available online: https://www.cobute.co.za/precast-decking-system (accessed on 22 August 2020).
186. Ferrante, C.D.O.; de Andrade, S.A.L.; de Lima, L.R.O.; Vellasco, P.C.G. Analytical study and experimental tests on innovative steel-concrete composite floorings. *J. Constr. Steel Res.* **2020**, *168*, 105868.
187. Lawson, M.; Beguin, P.; Obiala, R.; Braun, M. Slim-floor construction using hollow-core and composite decking systems. *Steel Constr.* **2015**, *8*, 85–89. [CrossRef]
188. Lu, X.; Mäkeläinen, P. Slim floor developments in Sweden and Finland. *Struct. Eng. Int. J. Int. Assoc. Bridg. Struct. Eng.* **1996**, *6*, 127–129. [CrossRef]
189. Braun, M.; Hechler, O.; Obiala, R. Untersuchungen zur Verbundwirkung von Betondübeln. *Stahlbau* **2014**, *83*, 302–308. [CrossRef]
190. Braun, M.; Obiala, R.; Odenbreit, C. Analyses of the loadbearing behaviour of deep-embedded concrete dowels, CoSFB. *Steel Constr.* **2015**, *8*, 167–173. [CrossRef]
191. Ju, Y.K.; Chun, S.C.; Kim, S.D. Flexural test of a composite beam using asymmetric steel section with web openings. *J. Struct. Eng.* **2009**, *135*, 448–458. [CrossRef]
192. Huo, B.Y.; D'Mello, C.A. Shear Transferring Mechanisms in a Composite Shallow Cellular Floor Beam with Web Openings. *Structures* **2017**, *9*, 134–146. [CrossRef]
193. Huo, B.Y.; D'Mello, C.A. Push-out tests and analytical study of shear transfer mechanisms in composite shallow cellular floor beams. *J. Constr. Steel Res.* **2013**, *88*, 191–205. [CrossRef]
194. Amran, Y.M.; El-Zeadani, M.; Lee, Y.H.; Lee, Y.Y.; Murali, G.; Feduik, R. Design innovation, efficiency and applications of structural insulated panels: A. review. *Structures* **2020**, *27*, 1358–1379. [CrossRef]
195. Lam, D.; Dai, X.; Kuhlmann, U.; Raichle, J.; Braun, M. Slim-floor construction—Design for ultimate limit state. *Steel Constr.* **2015**, *8*, 79–84. [CrossRef]

materials

MDPI

Article

Characterization and Vibro-Acoustic Modeling of Wood Composite Panels

Andrea Santoni [1,†,*], Paolo Bonfiglio [2,†], Patrizio Fausti [1,†], Cristina Marescotti [1,†], Valentina Mazzanti [1,†] and Francesco Pompoli [1,†]

1 Engineering Department, University of Ferrara, 44122 Ferrara, Italy; patrizio.fausti@unife.it (P.F.); cristina.marescotti@unife.it (C.M.); valentina.mazzanti@unife.it (V.M.); francesco.pompoli@unife.it (F.P.)

2 Materiacustica srl, 44122 Ferrara, Italy; paolo.bonfiglio@materiacustica.it

* Correspondence: andrea.santoni@unife.it

† These authors contributed equally to this work.

Received: 18 March 2020; Accepted: 14 April 2020; Published: 17 April 2020

check for updates

Abstract: Natural fiber-filled polymers offer good mechanical properties and economic competitiveness compared to traditional materials. Wood flour is one of the most widely used fillers, and the resulting material, known as wood plastic composite (WPC), has already found a wide applicability in many industrial sectors including automotive and building construction. This paper, as a followup of a previous study on a numerical-based approach to optimize the sound transmission loss of WPC panels, presents an extensive numerical and experimental vibro-acoustic analysis of an orthotropic panel made out of WPC boards. Both structural and acoustical excitations were considered. The panel radiation efficiency and its transmission loss were modeled using analytic and semi-analytic approaches. The mechanical properties of the structure, required as input data in the prediction models, were numerically determined in terms of wavenumbers by means of finite element simulations, and experimentally verified. The accuracy of the predicted acoustic performances was assessed by comparing the numerical results with the measured data. The comparisons highlighted a significant influence of the junctions between the WPC boards, especially on the panel's transmission loss. The radiation efficiency results were mostly influenced by the boundary conditions of the plate-like structure. This latter aspect was further investigated through a finite element analysis.

Keywords: wood plastic composite; transmission loss; radiation efficiency; orthotropic panel; wavenumber analysis

1. Introduction

The building construction industry is responsible for a significant amount of CO_2 global emissions and energy consumption. Taking climate actions within this sector can be very effective, even though, as shown in the 2019 Global Status Report on buildings and construction [1], the final energy demand in buildings is still rising. The strategic actions that should be undertaken in order to pursue the decarbonization of the building and construction industry span from increasing the use of renewable energy sources, to installing more efficient heating, cooling, and ventilation systems, but it also involves the development of innovative materials or bio-based solutions with a reduced impact on the environment. In order to be competitive against traditional systems, such innovative solutions need to be optimized both in terms of performances, as well as in their manufacturing process, other than being economically sustainable. The use of thermal and acoustical insulating systems involving natural or recycled materials is only at an early stage and still limited, as depicted by the state-of-the-art published a few years ago [2]. Nevertheless, natural fibers may represent a relatively inexpensive and valuable alternative to traditional synthetic thermal or sound insulation materials [3]. In fact, as was shown in a recently published paper [4], with an appropriate manufacturing process, it is possible to obtain natural hemp-based fibrous materials with a sound absorption coefficient comparable to the one provided by traditional synthetic fibrous materials. Besides, natural fiber-filled polymers offer both good mechanical properties and economic competitiveness compared to traditional materials. Among natural fibers, wood flour is one of the most widely used fillers, mainly because of its wider availability. The resulting material is usually known as wood plastic composite (WPC) [5]. The optimum properties are obtained using a wood fiber filling level of about 50 wt.%, the fibers possessing an aspect ratio of 10 or higher and using an amount of coupling agent around 4 wt.% [6,7]. As found in recent studies [8,9], the use of polypropylene as the matrix would improve the mechanical properties of WPC systems. However, it would also result in a more difficult processing and characterization [10,11]. In fact, the use of polyethylene allows for a lower processing temperature, reducing the risk of degrading the wood fiber [12]. WPC boards and panels are already widely involved in external flooring and decking, thanks to a better outdoor durability also in a wet environment compared to natural wood. WPC components also find wide application in the automotive industry [13]. Moreover, WPC systems can also be used in building construction applications [14] such as roofing systems, walls, and façades, or again as sound barriers. Due to the increasing interest in such a composite material, in the last few years, several studies have been published, investigating for example the influence of different wood fibers or of different types of matrix [15] or proposing the optimization and improvement of the mechanical or thermal properties of WPC systems [16,17].

As a followup to a previous study [18], which proposed a numerically-based approach to optimize the sound insulation performance of WPC panels, this paper presents an extensive numerical and experimental vibro-acoustic analysis of an analogous structure, increasing the knowledge on its dynamic response to a structural or acoustic excitation. The structural wavenumbers obtained from finite element (FE) simulations were used as input data for analytical and semi-analytical computational models to predict the panel radiation efficiency and its transmission loss (TL). The accuracy of the predicted results was assessed by comparing the numerical and measured data. The aim of this work is not limited to the presentation of computationally efficient approaches to investigate orthotropic plate-like structures, but especially to highlight and explain the discrepancies that are likely to be found between the predicted performance and the one observed in real structures. In the next section, the considered WPC element is described. In the same section, the numerical models implemented to characterize the elastic properties of the structure and to investigate its acoustic performance, both in terms of radiation efficiency and TL, are briefly introduced. Section 3 describes the experimental setup and the measurements performed on a WPC panel. Numerical results are presented, discussed, and validated, by comparison with experimental data, in Section 4.

2. Materials and Methods

The test panel was made of extruded WPC boards of high density polyethylene (HDPE) filled with 50 wt.% of wood fibers from pine sawdust, with a density $\rho = 1316$ kg/m^3, manufactured by Iperwood srl (Ferrara, Italy). The material's static elastic modulus, $E = 5.4$ GPa, was determined from the frequency response function (FRF) of a clamped-free homogeneous bar of the WPC material, excited by an impact hammer [18], by means of a resonant approach based on Oberst's beam method [19]. The geometry of the 22×145 mm^2 cross-section of the tested hollow boards is shown in Figure 1. These extruded boards were used to build a WPC panel, as clearly explained in the following part of the paper.

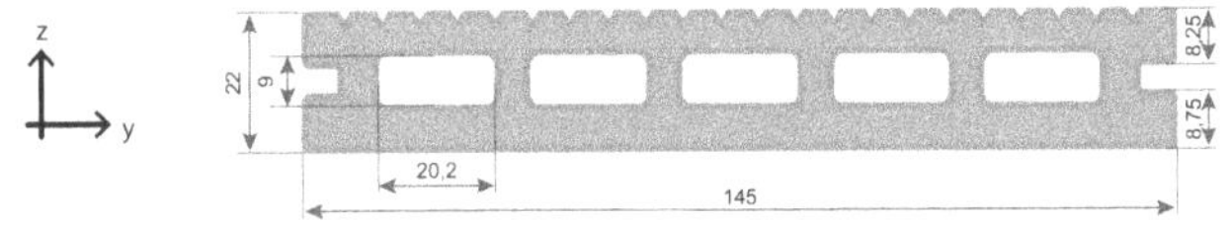

Figure 1. Cross-section of the tested wood plastic composite (WPC) boards. Dimensions are given in mm.

2.1. Characterization of the Panel's Elastic Properties

The bending stiffness of a plate-like structure is required as input data in several vibro-acoustic models. For the considered WPC components, this was investigated in terms of the wavenumbers of the structural waves, by means of a well-established wave-correlation approach. This required the evaluation of the dynamic response of the tested structure, usually due to a broadband excitation, along a line or over a grid of equally distributed points. This method was implemented in the form known as inhomogeneous wave correlation (IWC) [20–22]. A correlation function between a mono-dimensional inhomogeneous plane propagating wave, defined as $o\left(k_x\right) = \exp\left(-\mathrm{j}k_{x,r} + k_{x,i}\right)x$, and the vibrational field evaluated on the structure $\tilde{w}\left(\omega, x_i\right)$, can be expressed as:

$$\mathcal{F}\left(\omega, k_x\right) = \frac{\sum_i \tilde{w}\left(\omega, x_i\right) o^*\left(k_x\right) \Delta x}{\sqrt{\sum_n \left|\tilde{w}\left(\omega, x_i\right)\right|^2 \Delta x \sum_n \left|o^*\left(k_x\right)\right|^2 \Delta x}} \tag{1}$$

where ω is the angular frequency, Δx represents the spacing expressed in meters between two adjacent evaluation positions, while $k_{x,r}$ and $k_{x,i}$ represent the real and the imaginary components of the complex wavenumber k_x, the symbol * is used to indicate the complex conjugate, and j indicates the imaginary unit. The wavenumber dispersion relation is identified by maximizing the function $\mathcal{F}\left(\omega, k_x\right)$ given in Equation (1). The wave-correlation-based approaches allows investigating elastic and viscoelastic materials in a broad frequency range [23–26]. Moreover, it could be convenient to fit the evaluated wavenumbers with an analytical dispersion relation. In this case, the one derived form Mindlin's theory for thick plates [27] was used:

$$k^4 - \left[k_L^2 + \left(\frac{k_T}{\kappa}\right)^2\right] k^2 - k_B^4 + \frac{\left(k_L k_T\right)^2}{\kappa^2} = 0 \tag{2}$$

where k_L, k_T, and k_B represent the wavenumbers for longitudinal, transverse, and pure bending waves, respectively:

$$k_L = \omega\sqrt{\frac{\rho(1-\nu^2)}{E}}; \qquad k_T = \omega\sqrt{\frac{\rho}{G}}; \qquad k_B = \sqrt{\omega}\sqrt[4]{\frac{\mu}{D}} \tag{3}$$

The apparent elastic properties of the structure, E and G, are the fitting variable; ν represents the structure's Poisson ratio, D its effective bending stiffness, and ρ and μ its equivalent density and mass per unit of area, respectively. The coefficient κ, introduced to account for the non-uniform distribution of shear stress over the cross-section, can be computed from Poisson's ratio ν, as described in [28,29].

The dynamic response of the WPC components was initially investigated by modeling an extruded beam in an FE framework, by using the software COMSOL Multiphysics®, as shown in Figure 2a). WPC panels are constituted of beams assembled and joined together. Due to their geometric configuration, these panels may exhibit an orthotropic behavior. Therefore, it was necessary to also investigate the orthogonal direction, in which a plate would be made by laterally-jointed sections of extruded boards, as shown in Figure 2b). The junction between two adjoined boards was simplified by imposing a continuity condition at the interface surface. The 3D models for the considered beams were meshed ensuring that the maximum element size was ten times smaller than the wavelength associated with the highest investigated frequency. The beams were excited by a broadband boundary load applied in the vertical direction, as shown in Figure 3 by the velocity levels of the vertical component. The vibration field was evaluated over a line of points, evenly spaced 2 cm from one another.

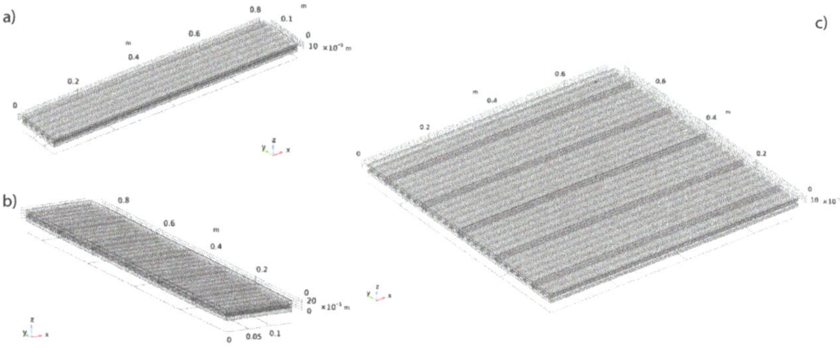

Figure 2. FE model of the tested WPC components: (**a**) x-wise wave propagation analysis: model of an extruded board; (**b**) y-wise wave propagation analysis: model of sections of boards coupled together; (**c**) model of the considered plate.

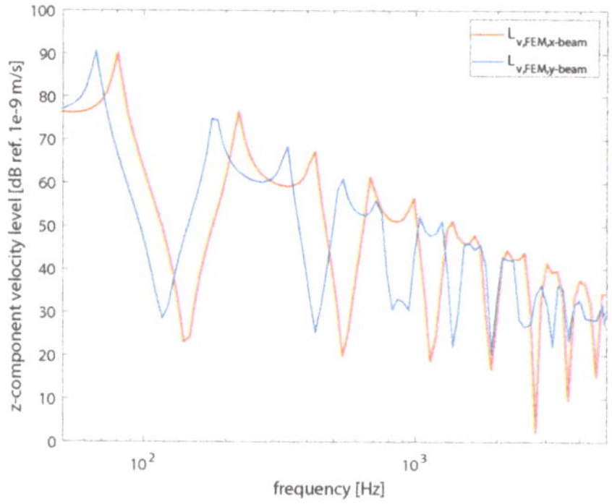

Figure 3. Vibration velocity levels of the vertical component of the vibration velocity computed from the FE models and averaged over the beams' surface.

2.2. Plate's Radiation Efficiency

The real part of the radiation impedance of a vibrating structure, normalized with respect to the acoustic impedance, represents its radiation efficiency σ. This is a non-dimensional vibro-acoustic descriptor that quantifies the capability of the structure to convert vibrational energy into acoustic energy as sound waves. It is particularly convenient to evaluate such a descriptor in order to assess the acoustic performance of a vibrating structure excited by mechanical sources. The radiation efficiency of the considered orthotropic WPC panel was computed by means of two different approaches. Firstly, the radiation efficiency was computed with a modal approach assuming the WPC panel as an equivalent thin orthotropic panel with simply supported boundaries and inserted into a infinite rigid baffle. The effect of fluid load was neglected. These very same assumptions on the plate configuration were also used to evaluate the modal-average radiation efficiency, considering a high modal density within the entire investigated frequency range. Both models take into account the orthotropic behavior of the WPC panel by means of a direction-dependent bending stiffness, which is defined as:

$$D(\phi) = D_x \cos^4 \phi + 2D_{xy} \cos^2 \phi \sin^2 \phi + D_y \sin^4 \phi \tag{4}$$

The angle ϕ represents the azimuthal angle of propagation of the structural wave. D_x and D_y are the bending stiffnesses associated with the plate's principal directions. D_{xy} is the effective torsional stiffness. In this case, the orthotropic bending stiffness given in Equation (4) was approximated by assuming $D_{xy} \approx \sqrt{D_x D_y}$. Thus, it was possible to compute an effective bending stiffness D_i along the ith principal direction from the wavenumbers k_i obtained from numerical models as described in the previous section:

$$D_i(\omega) = \frac{\omega^2 \mu_i}{k_i^4(\omega)}; \qquad i = x, y \tag{5}$$

The characterization of the elastic properties of the structure through the effective bending stiffness, given in Equation (5), allows compensating for the effects of shear deformation and rotatory inertia, neglected by the thin plate theory, which, however, may affect the plate's response at high frequencies. For all the details regarding the implemented equations, please refer to the work of Santoni et al. [30], in which the two models were developed in order to calculate the radiation efficiency of orthotropic cross-laminated timber panels.

It has already been proven that these two models are computationally efficient and provide accurate results for a thin orthotropic plate with simply supported boundaries. However, such a restraint condition is hardly feasible in real structures, which often present more complex boundary conditions. Therefore, we also investigated the radiation efficiency of the WPC panel by means of an FE model, using the software COMSOL Multiphysics®, allowing considering different boundary conditions. The plate, with a radiating surface $S = 0.73 \times 0.73 = 0.533$ m^2, was modeled coupling together five WPC boards by imposing a continuity condition at the interface surfaces. The mesh was generated in order to guarantee that the maximum element size was ten times smaller than the wavelength associated with the highest investigated frequency. A point force was applied in the position with coordinates $x_f = 0.65$ m and $y_f = 0.70$ m. The radiation efficiency was computed from the vibration velocity field evaluated over a uniform grid of points, spaced 5 cm from one another, by using a well-established hybrid approach know as the discrete calculation method [31]. This approach, assuming radiation in the free-field, combines the complex vibration velocity, evaluated over the surface of the considered structure, with the analytical computation of the radiation impedances. Therefore, the results obtained for the simply supported plate were directly comparable with the radiation efficiency analytically computed with the modal approach.

2.3. Plate's Transmission Loss

When a structure is excited by an acoustic field rather than a mechanical force, it is convenient to evaluate its acoustic performance in terms of sound transmission loss (TL). Sound transmission through the considered WPC panel was computed by means of the transfer matrix method (TMM) [32], an easily implementable approach with wide-ranging applicability to model wave propagation through laterally infinite media of different natures. The WPC plate was modeled as a single thin orthotropic layer [33,34], characterized by the wave impedance:

$$Z(\omega, \phi) = j\omega\mu \left(1 - \frac{D(\phi) k_t^4}{\omega^2 \mu}\right), \tag{6}$$

The orthotropic bending stiffness $D(\phi)$ is given in Equation (4), where k_t is the trace wavenumber defined as $k_t = k_0 \sin\theta$, with k_0 the acoustic wavenumber and θ the angle of incidence of the acoustic plane wave exciting the structure. The structure modeled within the TMM was assumed to be laterally of infinite extent; modal resonances and the diffraction effect caused by the finite dimension of real structures were not considered. Since sound transmission below the first coincidence is governed by forced vibration and modal resonances, such assumption may cause a significant lack of accuracy at low frequencies. However, the results could be improved by considering the non-resonant radiation efficiency of the finite size panel, as described by Villot et al. [35]. Other different formulations have been proposed to compute such a non-resonant radiation efficiency [36,37]; in particular in this case, the one developed by Rhazi et al. [38] was applied. Assuming a random incidence diffuse field excitation, the WPC plate's sound transmission loss is determined for each angular frequency ω as:

$$TL(\omega) = -10 \log \frac{\int_0^{2\pi} \int_0^{\pi/2} \tau(\omega, \theta, \phi) \sin\theta \cos\theta \, d\theta d\phi}{\int_0^{2\pi} \int_0^{\pi/2} \sin\theta \cos\theta \, d\theta d\phi} \tag{7}$$

where $\tau(\omega, \theta, \phi)$ is the sound transmission coefficient computed for each frequency ω with the TMM and eventually corrected with the non-resonant radiation efficiency of the finite size plate. It is also dependent on the angle of incidence of the impinging plane wave θ and the azimuthal angle ϕ.

3. Experimental Investigation

In order to validate the results computed from the prediction models and the numerical simulations described in the previous section, a WPC plate was experimentally investigated. A rectangular panel, with a radiating surface with the same dimensions considered in the numerical models: $L_x = 0.73$ m, $L_y = 0.73$ m and $h = 0.022$ m, was mounted into the testing window between the reverberant room and the semi-anechoic chamber of the University of Ferrara. The plate was realized joining together six WPC boards by means of high-density elastic putty, with a measured density of about $\rho_{putty} = 2400$ kg/m^3. The panel was secured by drilling holes in some of the boards and installing a wood frame fixed with threaded bolts. Since mechanical fixings were not employed to join the WPC beams, a significant amount of elastic putty was used to build the panel. It was applied both on the beams' junctions, through their entire thickness, and on the panel edges in order to keep the panel in place and especially to prevent any sound leakage. Figure 4 provides some photos of the experimental setup. Moreover, a digram of the junction between two WPC boards is shown in Figure 4e). Even though the presence of the elastic putty was neglected in the numerical model, by assuming a continuity condition between adjacent boards, it certainly influenced the wave propagation. Nevertheless, due to an obvious complexity and to the lack of knowledge of the material properties, it was not possible to quantify these effects accurately. However, as will be shown in the next section, it was possible to evaluate the effect of the additional mass, introduced by the elastic putty, on the panel dynamic response.

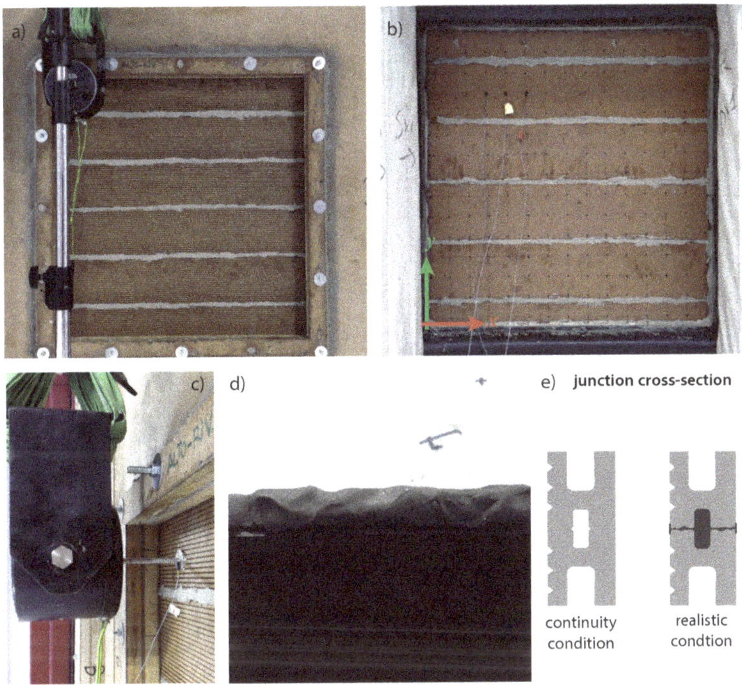

Figure 4. Pictures of the experimental setup used to investigate the WPC response to both a mechanical and an acoustic excitation: (**a**) panel view from the exciting side (reverberant room); (**b**) panel view from the receiving side (semi-anechoic room); (**c**) detail of the mechanical excitation; (**d**) details of the elastic putty layer applied to join the WPC boards vertically; (**e**) digram of the junction between two WPC boards.

The plate was mechanically excited by means of an electro-dynamic shaker B&K Type 4809 driven with a white noise signal. The shaker was suspended with rubber bands on the reverberant room side and rigidly fixed by gluing the stinger termination to the WPC plate, in the position $x_f = 0.65$ m and $y_f = 0.70$ m, according to the system shown in Figure 4, in order to replicate the numerical model. To evaluate the structural wavenumbers propagating along the two principal directions experimentally, the electro-dynamic shaker was driven with a broadband sine-sweep signal, from 50 Hz to 6000 Hz. The vibrational field was measured by means of miniature accelerometers PCB 352C22 with a sensitivity of 10 mV/g, along two orthogonal lines of points, evenly spaced 2 cm from one another, aligned with the excitation position. The impulse response of each measured point with respect to the reference signal was determined by means of the convolution technique. More details on this experimental approach can be found in [25]. Moreover, the vibration field on the panel's surface, due to a structure-borne white noise excitation within the frequency range 50 Hz–10 kHz, was also measured over a grid of points with a spacing of 5 cm, in order to evaluate the radiation efficiency of the WPC plate by means of the discrete calculation method, already introduced in the previous section. So as to have a constant phase relationship between non-simultaneous measurement, a reference transducer was placed on the shaker stinger termination. The vibration velocity was derived by dividing the measured acceleration by the factor $j\omega$, within the frequency domain.

In the reverberant room, which had a volume of 250.7 m^3, three omnidirectional sound sources were separately driven by a unrelated stationary white noise, and the average sound pressure level L_p was measured using six microphones placed in different positions. The receiving room had highly absorbing lateral walls and ceiling and a reflective floor. The average sound intensity L_i was measured inside this room at a distance of approximately 10-15 cm from the panel, by a manual scanning procedure using a B&K 3547 sound intensity probe. The plate's sound transmission loss was determined from such quantities according to the ISO 15186-1:2003 Standard [39].

4. Results

A first comparison was made between the wavenumbers evaluated from the FE beam models, introduced in Section 2, and the experimental data measured on the WPC plate, as described in the previous section. Figure 5a) presents the wavenumbers, along the principal directions x and y, directly determined by maximizing the inhomogeneous wave correlation function, given in Equation (1), and the fitted curve obtained using Mindlin's dispersion relation given in Equation (2). A rather good agreement was found between the numerical and experimental wavenumbers determined along the x-direction. On the other hand, above 2000 Hz, the experimental wavenumber measured along the y-direction was significantly higher than the wavenumber computed from the FE model. According to the results determined from the FE model, along the x-direction, the first coincidence frequency fell between 1100 Hz and 1400 Hz, consistent with what was found from the experimental wavenumbers. While along the y-direction, the first coincidence frequency was found around 1900 Hz from the numerical model, it shifted up to approximately 3000 Hz according to the experimental wavenumbers measured on the WPC plate. The increase of the structural wavenumber, which corresponded to a reduction of the bending stiffness, was caused by the mass added to the structure by the elastic putty, which was not considered in the numerical model. The presence of this material, used to join the WPC boards vertically, strongly affected the y-direction of propagation, while its influence was negligible along the x-direction. In fact, as shown in Figure 5b), the structural wavenumbers obtained along the y-direction from the FE model by considering the additional mass of the elastic putty were in rather good agreement with the experimental results.

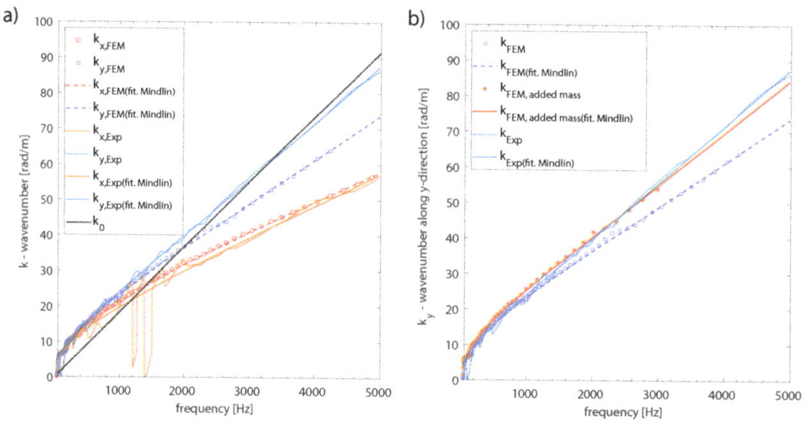

Figure 5. Wavenumbers determined numerically (FEM) and experimentally (Exp). (**a**) Comparison between the structural wavenumbers along the principal directions of the orthotropic WPC panel and the acoustic wavenumber; (**b**) wavenumbers determined along the y-direction considering the additional mass of the elastic putty.

It should be noted that the way in which the boards were coupled in order to realize a plate-like structure was suitable for laboratory conditions, but it would never be implemented in practical structures, for which the junctions need to be properly and conveniently designed. However, it is interesting to evaluate how the discrepancies observed between the numerical and the experimental wavenumbers, when the junctions were not considered, may affect the prediction of the acoustic performance of the structure. Therefore, the radiation efficiency and the transmission loss of the WPC panel computed using as input data the numerical wavenumbers, obtained from the FE models, were compared both with the results computed from the experimental wavenumbers and with the radiation efficiency and the TL measured in the laboratory. As shown in Figure 6, the TMM provided a good approximation of the experimental TL, except in the lowest frequency bands, were the panel response was governed by structural modes, which were not considered in the TMM framework. At the higher frequencies, using the experimental wavenumbers as input data certainly offered a better accuracy. In fact, compared to the results computed from the FE-based wavenumber, the coincidence region was better represented. The mass added to the system tended to reduce the stiffness above 2000 Hz. As already mentioned, this shifted the first coincidence towards higher frequencies. Moreover, it allowed better approximating the TL above the critical frequency. In fact, the experimental wavenumber increased with the frequency at a higher rate, with values that were close to the acoustic wavenumber in a wide range of frequencies. For this reason, the experimental wavenumbers allowed for a better prediction of the plate's TL, which exhibited a wider coincidence region. Figure 6 also reports the TL computed using the wavenumbers determined from the FE model in which the added mass was considered along the y-direction, obtaining a greater accuracy above the critical frequency.

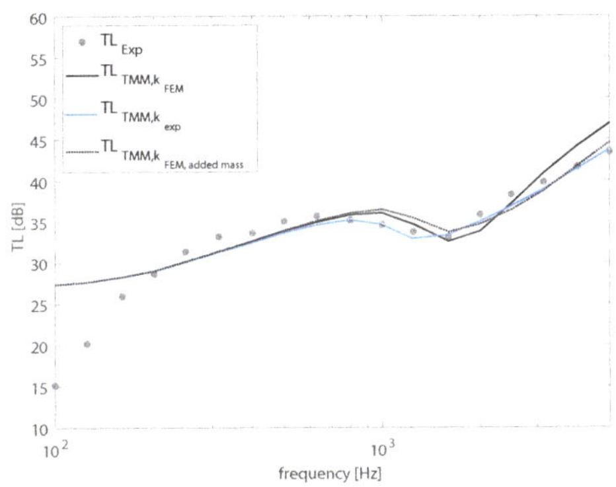

Figure 6. Transmission loss (TL) of the WPC sandwich plate in 1/3-octave bands. Comparison between the results computed by means of the transfer matrix method (TMM) using as input data the structural wavenumber TL_{TMM,k_j} and the experimental data TL_{Exp}.

An analogous comparison was made between the radiation efficiency measured in the laboratory and the modal-average radiation efficiency computed from numerical and experimental wavenumbers, as reported in Figure 7. Even though some of the effects described for the plate's TL also characterized sound radiation, the comparison in terms of radiation ratio $L_\sigma = 10\log\sigma$, given in Figure 7, highlighted more severe discrepancies between computed and numerical results. In fact, the effect of the added mass was limited to a wider and a less sharp peak associated with the coincidence region, obtained from the numerical wavenumbers. A slightly better accuracy could be obtained by taking into account the effect of the added mass in the numerical model employed to determine the structural wavenumbers, even though a significant deviation could still be observed. Thus, the effect of the mass added by the elastic putty along the junctions did not explain why above 2000 Hz, a substantially lower radiation ratio was experimentally determined. Both radiation efficiency models, introduced in Section 2, assumed simply supported boundary conditions. Although this assumption allowed for a computationally efficient model, such a condition is difficult to realize in practice. The boundary conditions of the tested WPC panel were rather complex and difficult to reproduce with a numerical model. However, it was possible to investigate their influence on sound radiation analyzing the radiation efficiency obtained by means of the DCMfrom the vibration velocity distribution evaluated using an FE model of the plate with several boundary conditions: SS-SS indicates that all the edges are simply supported (this condition is directly comparable with the modal-based analytic formulation); FF-FF indicates that all the edges are free; SS-FF indicates that the left and right edges are simply supported and the top and bottom edges are free; FF-SS indicates that the left and right edges are free and the top and bottom edges are simply supported.

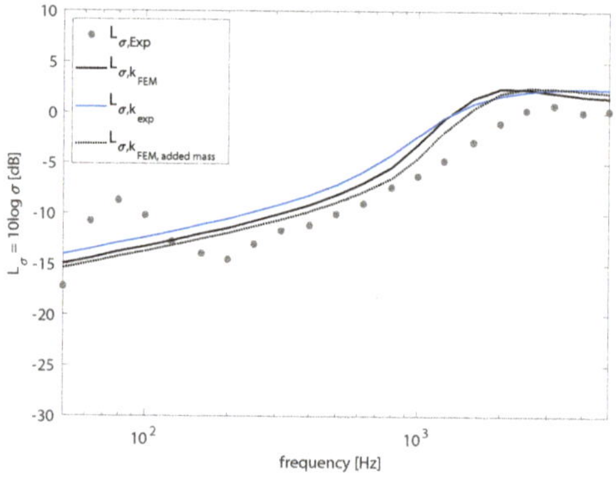

Figure 7. Radiation ratio of the WPC panel in 1/3-octave bands. Comparison between the results computed using as input data the structural wavenumber, determined from FE model $L_{\sigma,k_{FEM}}$, or experimentally $L_{\sigma,k_{exp}}$, and the experimental data $L_{\sigma,Exp}$.

As shown both in narrow frequencies and in one-third octave bands in Figure 8, the radiation ratio obtained from the FE model with simply supported boundary conditions $L_{\sigma,FEM,ss-ss}$ was in good agreement with the analytical model fed with the numerical structural wavenumbers as input data $L_{\sigma,k_{FEM}}$. The peak exhibited by the ration ratio at the lowest frequencies seemed to indicate that the simply supported condition was a good approximation of the edge fixing of the installed plate. However, as the frequency increased, the experimental radiation ratio was better approximated by the free-edge condition or by a combination of free and simply supported edges. In fact, as already found in other studies [40], decreasing the degree of restraint at the edges reduced the radiation efficiency of the panel. The restraint at the plate's edges was provided by a wood frame fixed with threaded bolts. Strips of a resilient viscoelastic material were interposed between the wood frame and the WPC panel along its perimeter, to prevent sound leakages. Results showed that, at low frequencies, the cross-section of the plate at the border was able to rotate, but the horizontal motion was prevented. However, as the frequency increased, the restraint to the horizontal motion provided by the experimental setup decreased, allowing, at least to some extent, the cross-section of the beam to move back and forth.

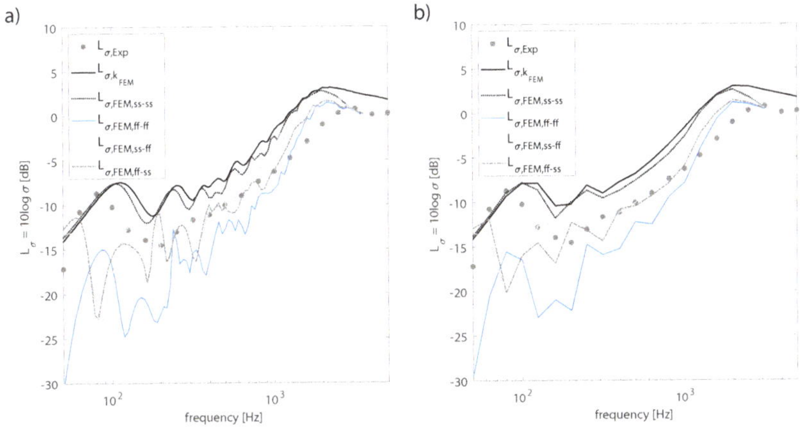

Figure 8. Influence of the boundary condition of the radiation ratio. Comparison between the modal-based analytical approach and the data obtained from FE simulations. (**a**) Radiation ratio narrow frequency bands; (**b**) radiation ratio 1/3-octave bands.

5. Conclusions

This paper presented a vibro-acoustic analysis of a WPC orthotropic panel, in which the predicted results were compared with measured data. The sound transmission loss and the radiation efficiency of a WPC panel were computed by means of analytical and semi-analytical models. The input data for these models were the structural wavenumbers associated with the plate's principal directions, obtained from FE simulations. A first validation, performed comparing the numerical wavenumbers with experimental results, highlighted a significant deviation along the direction over which the extruded boards were joined. In fact, the additional mass introduced onto the plate surface by the high-density elastic putty used to join the WPC boards strongly affected wave propagation along the vertical direction, emphasizing the orthotropic behavior of the panel. Even though the predicted TL approximated with satisfying accuracy the general trend of the experimental data, it was found that a better accuracy could be reached by using as input data the experimental TL, taking into account the effects of the added mass. These findings highlighted the importance of a proper design of the junction systems of real plate-like WPC structures; a convenient optimization of these joints might represent a possibility to further increase the vibro-acoustic performance of the system. The radiation efficiency obtained from analytical models was found to overestimate the experimental results. By means of an FE analysis, it was shown that the deviation between the predicted and measured radiation efficiency was caused by the real boundary conditions of the WPC panel, which were not simply supported as assumed by the analytic models. However, the employed analytical models represented a quick investigation tool for a preliminary evaluation of the radiation efficiency of the structure with simply supported boundaries, which perfectly matched the FE results. If required, it was possible to obtain a higher accuracy by means of numerical approaches, taking into account the real fixing condition of the plate's edges, when these could be modeled.

Author Contributions: Conceptualization, A.S., P.B., P.F. and F.P.; Data curation, A.S., C.M. and V.M.; Formal analysis, A.S., C.M. and V.M.; Investigation, A.S., C.M. and V.M.; Methodology, A.S., P.B., P.F. and F.P.; Project administration, P.F.; Resources, P.F. and F.P.; Software, P.B.; Supervision, P.F. and F.P.; Writing – original draft, A.S.; Writing – review and editing, P.B., P.F., C.M., V.M. and F.P. All authors have read and agreed to the published version of the manuscript.

Funding: This research received no external funding.

Acknowledgments: The authors would like to thank Andrea Pizzardi and the company Iperwood for providing the WPC boards.

Conflicts of Interest: The authors declare no conflict of interest.

References

1. Global Alliance for Buildings and Construction, International Energy Agency and United Nations Environment Programme, *2019 Global Status Report for Buildings and Construction: Towards a zero-emissions, efficient and resilient buildings and construction sector.* International Energy Agency, Paris, France, 2019.
2. Asdrubali, F.; D'Alessandro, F.; Schiavoni, S. A review of unconventional sustainable building insulation materials. *Sustain. Mater. Technol.* **2015**, *4*, 1–17.
3. Asdrubali, F.; Schiavoni, S.; Horoshenkov, K. A review of sustainable materials for acoustic applications. *Build. Acoust.* **2012**, *19*, 283–311.
4. Santoni, A.; Bonfiglio, P.; Fausti, P.; Marescotti, C.; Mazzanti, V.; Mollica, F.; Pompoli, F. Improving the sound absorption performance of sustainable thermal insulation materials: Natural hemp fibers. *Appl. Acoust.* **2019**, *150*, 279–289.
5. Klyosov, A.A. *Wood-Plastic Composites*; John Wiley & Sons: Hooboken, NJ, USA, 2007.
6. Lu, J.Z.; Wu, Q.; Negulescu, I.I. Wood-fiber/high-density-polyethylene composites: Coupling agent performance. *Appl. Polymer Sci.* **2005**, *96*, 93–102.
7. Schirp, A.; Stender, J. Properties of extruded wood-plastic composites based on refiner wood fibers (TMP fibers) and hemp fibers. *Eur. J. Wood Wood Prod.* **2010**, *68*, 219–231.
8. Englund, K.; Villechevrolle, V. Flexure and water sorption properties of wood thermoplastic composites made with polymer blends. *Appl. Polymer Sci.* **2011**, *120*, 1034–1039.
9. Mazzanti, V.; Mollica, F.; El Kissi, N. Rheological and mechanical characterization of polypropylene-basedwood plastic composites. *Polym. Compos.* **2016**, *37*, 3460–3473.
10. Mazzanti, V.; Mollica, F. In-line rheometry of polypropylene based wood polymer composites. *Polym. Test.* **2015**, *47*, 30–35.
11. Mazzanti, V.; Mollica, F. Pressure dependent wall slip of wood flour filled polymer melts. *Journal of Non-Newtonian Fluid Mechanics* **2017**, *247*, 178–187.
12. Migneault, S.; Koubaa, A.; Erchiqui, F.; Chaala, A.; Englund, K.; Wolcott, M.P. Effects of processing method and fiber size on the structure and properties of wood–plastic composites. *Compos. Part A Appl. Sci. Manuf.* **2009**, *40*, 80–85.
13. Ashori, A. Wood–plastic composites as promising green-composites for automotive industries! *Bioresour. Technol.* **2008**, *99*, 4661–4667.
14. Fan, M.; Fu, F. Introduction: A perspective–natural fiber composites in construction. In *Advanced High Strength Natural Fibre Composites in Construction*; Elsevier: Amsterdam, The Netherlands, 2017; pp. 1–20.
15. Taufiq, M.; Mansor, M.R.; Mustafa, Z. Characterisation of wood plastic composite manufactured from kenaf fiber reinforced recycled-unused plastic blend. *Compos. Struct.* **2018**, *189*, 510–515.
16. Zhang, Y.; Xue, P.; Ding, Y.; Jia, M.; Cai, J.; Jin, X. Improvement of mechanical properties of wood-plastic composite floors based on the optimum structural design. *Acta Mech. Solida Sinica* **2016**, *29*, 444–454.
17. Qi, C.; Zhang, F.; Mu, J.; Zhang, Y.; Yu, Z. Enhanced mechanical and thermal properties of hollow wood composites filled with phase-change material. *J. Cleaner Prod.* **2020**, *256*, 120373.
18. Santoni, A.; Bonfiglio, P.; Mollica, F.; Fausti, P.; Pompoli, F.; Mazzanti, V. Vibro-acoustic optimization of wood plastic composite systems. *Constr. Build. Mater.* **2018**, *174*, 730–740.
19. ASTM E756-05–Standard Test Method for Measuring Vibration-Damping Properties of Materials; Standard; ASTM International: West Conshohocken, PA, USA, 2010.
20. Berthaut, J.; Ichchou, M.; Jezequel, L. K-space identification of apparent structural behavior. *J. Sound Vib.* **2005**, *280*, 1125–1131.
21. Ichchou, M.; Berthaut, J.; Collet, M. Multi-mode wave propagation in ribbed plates: Part I, wavenumber-space characteristics. *Int. J. Solids Struct.* **2008**, *45*, 1179–1195.
22. Ichchou, M.N.; Berthaut, J.; Collet, M. Multi-mode wave propagation in ribbed plates. Part II: Predictions and comparisons. *Int. J. Solids Struct.* **2008**, *45*, 1196–1216.
23. Van Damme, B.; Zemp, A. Measuring Dispersion Curves for Bending Waves in Beams: A Comparison of Spatial Fourier Transform and Inhomogeneous Wave Correlation. *Acta Acust. United Acust.* **2018**, *114*, 228–234.

24. Roozen, N.B.; Labelle, L.; Leclere, Q.; Ege, K.; Alvarado, S. Non-contact experimental assessment of apparent dynamic stiffness of constrained-layer damping sandwich plates in a broad frequency range using a Nd: YAG pump laser and a laser Doppler vibrometer. *J. Sound Vib.* **2017**, *395*, 90–101.
25. Santoni, A.; Bonfiglio, P.; Fausti, P.; Pompoli, F. Alternative method to the Oberst technique to measure the complex elastic modulus of visco-elastic materials. *Noise Control Eng. J.* **2019**, *67*, 1–10.
26. Tufano, G.; Errico, F.; Robin, O.; Droz, C.; Ichchou, M.; Pluymers, B.; Desmet, W.; Atalla, N. K-space analysis of complex large-scale meta-structures using the Inhomogeneous Wave Correlation method. *Mech. Syst. Signal Process.* **2020**, *135*, 106407.
27. Nilsson, A.; Liu, B. *Vibro-Acoustics*; Springer: Berlin/Heidelberg, Germany, 2016; Volume 2.
28. Graff, K.F. *Wave Motion in Elastic Solids*; Dover Publications, Inc.: New York, NY, USA, 1991.
29. Santoni, A.; Schoenwald, S.; Van Damme, B.; Fausti, P. Determination of the elastic and stiffness characteristics of cross-laminated timber plates from flexural wave velocity measurements. *J. Sound Vib.* **2017**, *400*, 387–401.
30. Santoni, A.; Schoenwald, S.; Fausti, P.; Tröbs, H.M. Modelling the radiation efficiency of orthotropic cross-laminated timber plates with simply-supported boundaries. *Appl. Acoust.* **2019**, *143*, 112–124.
31. Hashimoto, N. Measurement of sound radiation efficiency by the discrete calculation method. *Appl. Acoust.* **2001**, *62*, 429–446.
32. Allard, J.; Atalla, N. *Propagation of Sound in Porous Media: Modelling Sound Absorbing Materials*, 2nd ed.; John Wiley & Sons, Ltd: Chichester, UK, 2009.
33. Atalla, N. Modeling the sound transmission through complex structures with attached noise control materials. *Wave Motion* **2014**, *51*, 650–663.
34. Santoni, A.; Bonfiglio, P.; Fausti, P.; Schoenwald, S. Predicting sound radiation efficiency and sound transmission loss of orthotropic cross-laminated timber panels. In Proceedings of the Meetings on Acoustics 173EAA, Boston, USA, 25-29 June 2017; Volume 30, p. 015013.
35. Villot, M.; Guigou-Carter, C.; Gagliardini, L. Predicting the acoustical radiation of finite size multi-layered structures by applying spatial windowing on infinite structures. *J. Sound Vib.* **2001**, *245*, 433–455.
36. Vigran, T.E. Predicting the sound reduction index of finite size specimen by a simplified spatial windowing technique. *J. Sound Vib.* **2009**, *325*, 507–512.
37. Bonfiglio, P.; Pompoli, F.; Lionti, R. A reduced-order integral formulation to account for the finite size effect of isotropic square panels using the transfer matrix method. *J. Acoust. Soc. Am.* **2016**, *139*, 1773–1783.
38. Rhazi, D.; Atalla, N. Transfer matrix modeling of the vibroacoustic response of multi-materials structures under mechanical excitation. *J. Sound Vib.* **2010**, *329*, 2532–2546.
39. ISO 15186-1–Acoustics–Measurement of Sound Insulation in Buildings and of Building Elements Using Sound Intensity–Part 3: Laboratory Measurements; Standard; International Organization for Standardization: Geneva, Switzerland, 2003.
40. Squicciarini, G.; Thompson, D.; Corradi, R. The effect of different combinations of boundary conditions on the average radiation efficiency of rectangular plates. *J. Sound Vib.* **2014**, *333*, 3931–3948.

materials

MDPI

Article

Sound Transmission Loss of a Sandwich Plate with Adjustable Core Layer Thickness

Tom Ehrig [1,*], Martin Dannemann [1], Ron Luft [1], Christian Adams [2], Niels Modler [1] and Pawel Kostka [1]

[1] Institute of Lightweight Engineering and Polymer Technology, Technische Universität Dresden, 01069 Dresden, Germany; martin.dannemann@tu-dresden.de (M.D.); ron.luft@tu-dresden.de (R.L.); niels.modler@tu-dresden.de (N.M.); pawel.kostka@tu-dresden.de (P.K.)

[2] Department of Mechanical Engineering, System Reliability, Adaptive Structures, and Machine Acoustics SAM, Technical University of Darmstadt, 64289 Darmstadt, Germany; christian.adams@sam.tu-darmstadt.de

* Correspondence: tom.ehrig@tu-dresden.de

Received: 27 August 2020; Accepted: 17 September 2020; Published: 18 September 2020

check for updates

Abstract: Compressible Constrained Layer Damping (CCLD) is a novel, semi-active, lightweight-compatible solution for vibration mitigation based on the well-known constrained layer damping principle. The sandwich-like CCLD set-up consists of a base structure, a constraining plate, and a compressible open-cell foam core in between, enabling the adjustment of the structure's vibration behaviour by changing the core compression using different actuation pressures. The aim of the contribution is to show to what degree, and in which frequency range the acoustic behaviour can be tuned using CCLD. Therefore, the sound transmission loss (TL), as an important vibro-acoustic index, is determined in an acoustic window test stand at different actuation pressures covering a frequency range from 0.5 to 5 kHz. The different actuation pressures applied cause a variation of the core layer thickness (from 0.9 d_0 to 0.3 d_0), but the resulting changes of the stiffness and damping of the overall structure have no significant influence on the TL up to approximately 1 kHz for the analysed CCLD design. Between 1 kHz and 5 kHz, however, the TL can be influenced considerably well by the actuation pressure applied, due to a damping-dominated behaviour around the critical frequency.

Keywords: sound transmission loss; semi-active damping; sandwich panel; morphing structure; compressible constrained layer damping; composite materials

1. Introduction

Fibre-reinforced plastics (FRP) are increasingly used in everyday products. In particular, the use of FRP with glass and carbon fibres as reinforcing materials has spread from the aerospace sector, via automotive engineering and medical technology, to general mechanical engineering due to their excellent mechanical properties at low weight [1]. Considering that mass, stiffness and damping are coupled design variables of FRP, the lightweight oriented design often causes a problematic vibration susceptibility of the developed components. The consequences of component vibrations range from the subjective feeling of a low-quality component, to actual noise exposure to the failure of the component under high dynamic loads. In order to avoid potential environmental and health issues as well as damage-relevant or function-impairing amplitudes caused by vibrations, systems for health monitoring [2,3] and solutions for vibration mitigation were developed. The latter can be passive, semi-active or active damping treatments [4–13]. Most of these damping concepts achieve very good results. However, the overall implementation costs of semi-active, and especially active, damping systems are still high. Moreover, the additional mass of structure-integrated and peripheral hardware is a major drawback for many applications [14]. As part of their current research, the authors are pursuing a novel approach to combine the two requirements—lightweight design and efficient adjustable damping—for a new generation of lightweight structures.

1.1. Operating Principle of Compressible Constrained Layer Damping

For the adaptation of the dynamic behaviour, the authors have already proposed an original concept of Compressible Constrained Layer Damping (CCLD) in previous publications [15–18]. CCLD is an extension of the well-known Constrained Layer Damping (CLD) design, where the usually incompressible viscoelastic damping layer, is replaced by a compressible one, which is able to undergo compression and expansion in thickness direction. The adaptation principle of CCLD is therefore based on the pressure-controlled thickness variation of the viscoelastic core layer (Figure 1). While the storage and loss shear modulus of the viscoelastic layer G′ and G′′, which affect the damping performance significantly, as well as the thickness of the viscoelastic layer are constant in CLD, they depend on the actuation pressure p in the case of CCLD.

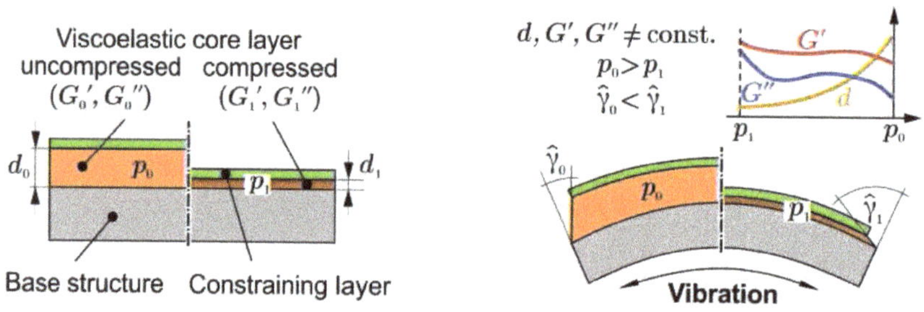

(**a**) (**b**)

Figure 1. Illustration of the Compressible Constrained Layer Damping (CCLD) principle: (**a**) set-up without bending; (**b**) under vibration.

The adjustable viscoelastic properties of the core layer, as well as the altering of the shear deformation amplitude, $\hat{\gamma}$, by variation of the core layer thickness, d, finally lead to a simple yet efficient adaptation of the dynamic behaviour of the overall system. The actual damping mechanism is based, as in CLD, on the shear deformation induced within the constrained damping layer. The actuation pressure could be provided, e.g., by hydraulic fluid, compressed air, or vacuum. However, in the present paper, only vacuum actuation will be considered.

In a previous publication [15], the authors have analysed the mobility (ratio of vibration velocity to excitation force) of a composite plate with CCLD in a lower acoustical frequency range. In this case of force excitation, which is relevant, e.g., in mechanical engineering, a complex, frequency-dependent control of the CCLD can significantly reduce the surface velocities, and thus the radiated sound power. In the present contribution, the sound absorbing ability of a CCLD-equipped plate is investigated.

1.2. Sound Transmission Loss of Sandwich Structures

The sound transmission loss (TL) is an important vibro-acoustic index to determine how well structures insulate sound. The TL is defined as the ratio of incident to transmitted sound power. There are a number of methods and publications for calculating the TL of FRP and sandwich structures, which are comprehensively reviewed by D'Alessandro et al. [19] and by Isaac et al. [20]. The studies of Dym and Lang [21] are of particular interest, as they describe the effects of skin thickness variation (and their ratio) and core stiffness variation in great detail. All of these models and experimental data show that the influence of the core layer properties (material, thickness, loss factor) on the TL is strongly dependent on the frequency range, especially if it is below or above the critical frequency. The influence of the foam core's shear stiffness on the critical frequency is modelled numerically in [22].

According to these calculations, a decrease of the critical frequency is expected with an increasing shear stiffness of the foam core. From the authors' own investigations [23], it is known that the correlation between the shear property changes of the foam core material and increasing compression is strongly non-linear. Furthermore, there is no linear correlation between skin-to-core thickness ratio and TL [24]. According to the work of Liu et al. [25,26], it can be expected that the TL improves with decreasing internal pressure. However, the publication of Du et al. [27] indicates that the variation of thickness and shear modulus of the viscoelastic core layer has no significant influence on the TL. All these results lead to the conclusion that it is hard to estimate the TL and the critical frequency as functions of actuation pressure in advance, although it is to be expected that the damping effect of the CCLD will be mainly present in the stiffness-dominated region, as well as in the damping-dominated region around and above the critical frequency.

The aim of the contribution is the experimental investigation of the acoustic effects of the CCLD concept. In particular, the change of TL and the shift of the critical frequency through varying the actuation pressure will be considered. The unique feature of the CCLD treatment in the acoustical context is that the geometric dimensions of the test structure remain constant (except for the core layer thickness d), as does the total mass of the structure. Simultaneously, the shear deformation kinematics on the structural level are changed by adjusting the actuation pressure, as well as damping and stiffness, of the viscoelastic core material, resulting in altered acoustic properties.

2. Materials and Test Set-up Configuration

2.1. Materials

The sandwich-like multilayer CCLD set-up consists of three layers: the base structure; the constraining layer; and the constrained viscoelastic core layer. It is an asymmetric sandwich structure, as the top layer is only one tenth of the thickness of the base structure, and, in addition, has high shear and tensile but low flexural stiffness.

A carbon fibre-reinforced plastics (CFRP) plate with a length of $a_b = 1100$ mm, a width of $b_b =$ 700 mm, and a thickness of $d_b = 2$ mm was used as base structure. The plate was made of seven prepreg layers (Co. Carboplast, Werne, Germany, HTA fibres and epoxy resin, hot-pressing process) with a [0/90,0_2,90,0_2,0/90] lay-up resulting in low intrinsic structural damping. The viscoelastic core layer consisted of the open-cell polyurethane foam Confor CF-47M (Co. Aearo Technologies LLC, Indianapolis, United States of America). The size of the viscoelastic core layer was set to be 880 mm × 580 mm to ensure complete CCLD coverage of the sound-exposed area by overlapping the dimensions of the acoustic window (860 mm × 560 mm). The thickness d_0 in the uncompressed state was equal to 10 mm. A CFRP sheet (Co. R&G, Waldenbuch, Germany) with the dimensions of 880 mm × 580 mm and a thickness of 0.2 mm was used as constraining layer. The CFRP sheet, which is produced by the technique of hot-pressing, consists of just one layer of HT carbon fibre prepreg (grammage 200 g/m^2, twill 2/2 weave) and an epoxy matrix. Table 1 contains the materials used and their main properties.

Table 1. Compressible Constrained Layer Damping (CCLD) materials and their main properties.

Layer	Material	Manufacturer	Density ρ	Length a	Width b	Thickness d	Mass m	Mass per Unit Area
			kg/m^3	mm	mm	mm	kg	kg/m^2
Base structure	CFRP (HTA-CF, EP)	Carboplast	1550	1100	700	2	2.41	3.13
Visco-elastic layer	Viscoelastic foam Confor CF-47M	Aearo Technologies LLC	96	880	580	10	0.56	1.10
Constrain-ing layer	CFRP (HT-CF, EP)	R&G	1550	880	580	0.2	0.18	0.35

2.2. CCLD Test Set-up Configuration

For the acoustic measurements, the CCLD treatment was set up as shown in Figure 2. The base structure, the damping layer, and the constraining layer were stacked on top of each other, while vacuum-sealing was assured by using a vacuum film and vacuum sealing tape. The individual layers were not glued together to avoid introducing another viscoelastic layer and to eliminate further phenomena, such as curing of the adhesive in the open-cell pores. The set-up was kept in position by the tightly applied vacuum film, and later fixed by the vacuum pressure enabling coupling of the layers through friction forces. The applied actuation pressure was generated by a vacuum pump and adjusted with a valve to achieve the predefined core layer thicknesses (Table 2). After setting the desired actuation pressure and a waiting time of >1 min an evenly distributed foam compression was achieved over the entire surface of the plate. Only at $p_1 = -16$ kPa (corresponding to 0.3 d_0) a slightly smaller compression of the foam was observed in the edge area up to max. 20 mm from the edges. The base structure with the applied CCLD treatment (hereinafter referred to as 'test structure' for the sake of simplicity) was then rigidly clamped in the acoustic window test stand using screws on all of its edges with the CCLD treatment facing the receiver room. The size of the base structure and the dimensions of the CCLD treatment were carefully chosen so that there were no air gaps or additional acousto-mechanical bridges in the test structure.

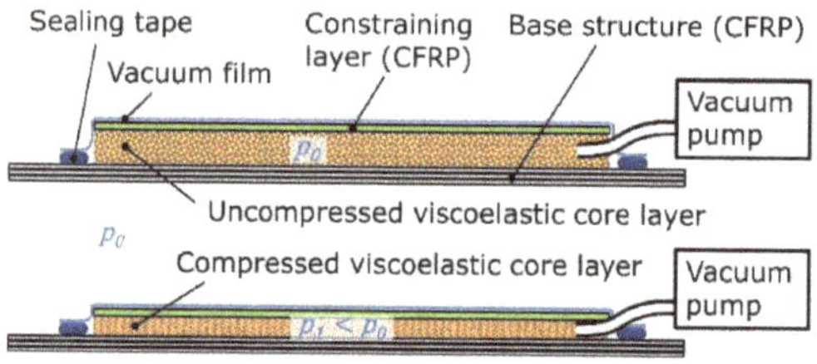

(a)

(b)

Figure 2. Lay-up of the base structure with CCLD treatment: (**a**) without actuation pressure (**top**) and with applied actuation pressure (**bottom**); (**b**) Test structure in the window test stand, receiver room.

Table 2. Applied actuation pressure and corresponding core layer thickness (d_0 = 10 mm).

Applied Actuation Pressure p_1	−3.9 kPa	−5.9 kPa	−7.6 kPa	−16.0 kPa
Core layer thickness d_1	0.9 d_0	0.7 d_0	0.5 d_0	0.3 d_0

2.3. Sound Transmission Loss Measurements

The TL was determined in an acoustic window test stand consisting of a transmitter room and a receiver room. Both rooms were connected by a window with the dimensions of 860 mm × 560 mm, which contained the test structure (Figure 3).

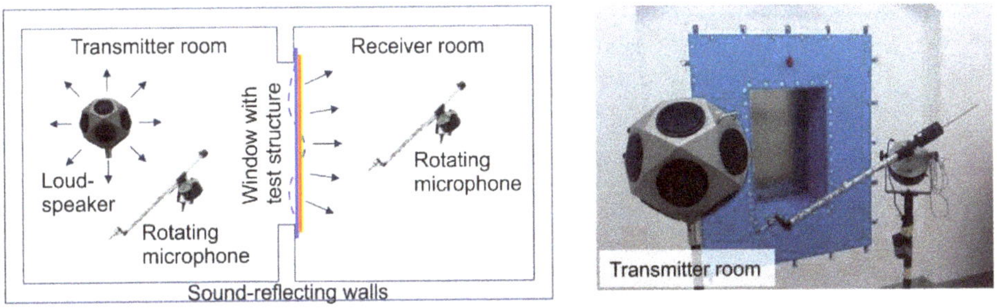

(**a**) (**b**)

Figure 3. (**a**) Set-up for the sound transmission loss measurement; (**b**) transmitter room of the acoustic test stand.

A diffuse sound field in the transmitter room was used to acoustically excite the test structure, which radiated sound into the receiver room. During excitation, the spatially and temporally averaged sound pressure was measured in the transmitter as well as in the receiver room using rotating microphones. The same set of microphone measuring positions was used for each applied actuation pressure. Based on those sound pressure measurements, the TL was calculated using the standardized procedure [28]. In accordance with the standards [28,29] and in order to ensure that a diffuse sound field was established within the given room dimensions, the lower limit of the measuring range was set at 0.5 kHz. According to standard [29], the upper limit of the measuring range is 3.15 kHz. Since our analytical calculation (see Section 4.1.) predicted an f_c at about 4.8 kHz, we chose the extended measuring range defined in [29], which specifies 5 kHz as the upper limit of the measuring range.

3. Sound Transmission Loss Calculation of the Base Structure

The TL of an acoustic wave that incidents an infinite plate under the angle ϕ with respect to the normal direction of a plate's surface can be written as [30]

$$\mathrm{TL} = 10\log_{10}\left(\frac{\left(\frac{2\rho_L c_L}{\cos(\phi)} + \frac{B_b}{2\pi f}\eta_b\left(\frac{2\pi f}{c_L}\sin(\phi)\right)^4\right)^2 + \left(2\pi f \rho_b d_b - \frac{B_b}{2\pi f}\left(\frac{2\pi f}{c_L}\sin(\phi)\right)^4\right)^2}{\left(\frac{2\rho_L c_L}{\cos(\phi)}\right)^2}\right)\mathrm{dB}, \quad (1)$$

where ρ_L is the air density, c_L is the speed of sound, η_b is the base structure's material loss factor, ρ_b is the mass density of the base structure, and f is the frequency. According to [31], the equivalent flexural stiffness of an orthotropic plate yields

$$B_b = \left(\frac{\sqrt{B_{b,0^\circ}}}{2} + \frac{\sqrt{B_{b,90^\circ}}}{2}\right)^2, \quad (2)$$

With

$$B_{b,0^\circ} = \frac{E_{b,0^\circ} d_b^3}{12\left(1-\mu_b^2\right)} \quad \text{and} \quad B_{b,90^\circ} = \frac{E_{b,90^\circ} d_b^3}{12\left(1-\mu_b^2\right)}. \quad (3)$$

$E_{b,0^\circ}$ and $E_{b,90^\circ}$ denote the Young's moduli in the plate's 0° direction and 90° direction, respectively, and μ_b denotes Poisson's ratio. Although Equation (1) assumes that the plate is infinite, its results are compared to the TL measurements of the base structure, which has finite dimensions. It is a first starting point to check and better understand the measurement results. Equation (1) is easy to implement, which is an advantage over more complex models such as finite element models. An improved model for TL calculations should take into account the boundary conditions of the base structure and the acoustic properties of the transmitter room and the receiver room. These aspects are neglected in Equation (1) and can be considered in numerical models only, which are beyond the scope of this paper. It is well known from the literature [14,19,21] that the TL can be divided into three characteristic frequency ranges. At frequencies up to $2f_r$, the TL is stiffness-dominated, where f_r denotes the fundamental frequency. From $2f_r$ through $f_c/2$ the TL is mass-dominated, where f_c denotes the critical frequency. Above the critical frequency the TL is damping-dominated. The following equations are used in this paper to estimate these frequency ranges for the base structure. As the base structure was clamped on all edges, its fundamental frequency could be estimated by [32]

$$f_r \approx \frac{6}{\pi}\sqrt{\frac{7}{2}\left(\frac{1}{a_b^4}+\frac{4}{7}\frac{1}{a_b^2 b_b^2}+\frac{1}{b_b^4}\right)}\sqrt{\frac{B_b}{\rho_b d_b}}. \quad (4)$$

The critical frequency f_c of the base structure reads [32]

$$f_c = \frac{c_L^2}{2\pi}\sqrt{\frac{\rho d_b}{B_b}}. \quad (5)$$

4. Results and Discussion

This section first characterizes the base structure and compares the measured results to those from TL calculations of an infinite plate. Secondly, the results of the test structure are discussed.

4.1. Base Structure

With the speed of sound $c_L = 343$ m/s, the air's density $\rho_L = 1.2$ kg/m^3, and the base structure's parameters ($a_b = 1100$ mm, $b_b = 700$ mm, $d_b = 2$ mm, $\rho_b = 1550$ kg/m^3, $\mu_b = 0.35$, $E_{b,0^\circ} = 62$ GPa, and $E_{b,90^\circ} = 60$ GPa [33]) the critical frequency yields 4.85 kHz according to Equation (5), and the fundamental frequency yields 33.3 Hz according to Equation (4). The mass-dominated region ranges from $2f_r = 66.6$ Hz through $f_c/2 = 2.42$ kHz. Figure 4 shows the measured TL of the base structure (for more detail see also Figure 5) as well as the calculation results for three different incidence angles using Equation (1). Between 0.5 kHz and 0.8 kHz, the measured TL agrees well with the TL of an infinite plate if the sound waves incident in normal direction ($\phi = 0°$). Between 1 kHz and 2 kHz, the measured TL increases by approximately 3 dB, which is half of the increase that would have been expected by the mass law. Between $f_c/2 = 2.42$ kHz and $f_c = 4.85$ kHz, the measured TL differs from the calculated one ($\phi = 0°$) to a higher extent. In order to check whether the results in this frequency range can also be described by Equation (1), the incidence angle was iteratively increased until the calculated TL matched the measured one. At an angle of $\phi = 60°$ the TL measurement and the TL calculation agree sufficiently well, see Figure 4. Thus, between $f_c/2$ and f_c the measured TL equates more to that of an infinite plate that is excited by an acoustic wave under an angle of $\phi = 60°$ than the TL of an infinite plate, where the acoustic waves incident in normal direction. In other words, the (finite) base structure behaves similarly to an infinite plate, where the acoustic wave incidents with an angle of $\phi = 60°$. These results suggest that all the aspects neglected in Equation (1), such as the base structure's boundary conditions or the acoustic properties of transmitter room and receiver room, can be condensed into the incident angle ϕ. As such, the incident angle ϕ in Equation (1) could rather be seen as a hypothetical incident angle for an infinite plate than as the actual incident angle during the measurements. The loss factor of the base structure η_b also affects the TL. As can be seen from Figure 4, this only affects the dip of the TL above the critical frequency. Since this dip is out of the measured frequency range, the loss factor of the base structure does not significantly affect the TL measurements performed in this paper. Consequently, the simple model of the TL shown in Equation (1) can qualitatively explain the measured TL if the incident angle ϕ is considered a parameter that covers all simplifications of the model with respect to the real measurement setup. Thus, the measured TL of the base structure is considered validated according to literature [14,19,21].

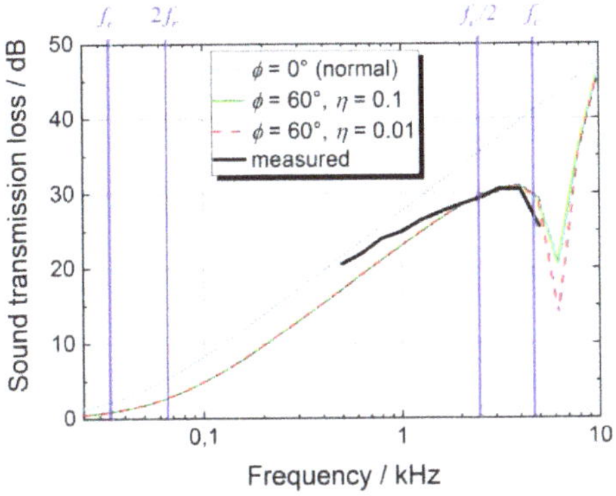

Figure 4. Measured and calculated sound transmission loss of the base structure versus frequency; vertical lines indicate the labelled frequencies.

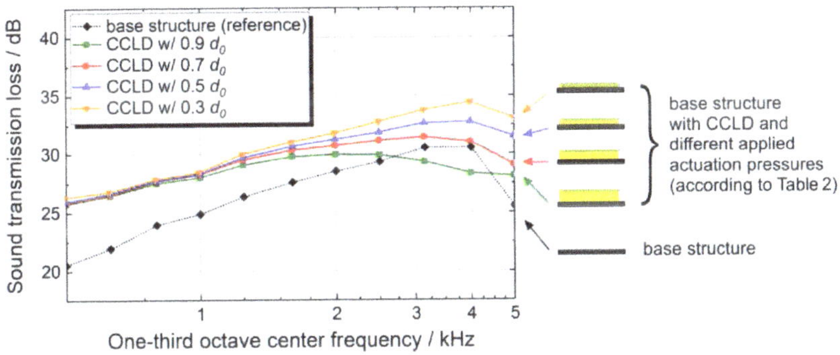

Figure 5. Sound transmission loss of the investigated test structure (base structure with CCLD treatment) at various core layer thicknesses (as result of different applied actuation pressures) compared to the transmission loss (TL) of the bare base structure. The repeat accuracy is <0.35 dB over the entire frequency range, therefore no error bars are displayed.

4.2. Base Structure with CCLD Treatment

Figure 5 shows the TL of the base structure as well as the TL of the test structure with various thicknesses. The frequency range from 0.5 kHz up to 1 kHz lies within the mass-controlled region, which is why there are no significant TL differences between the various compression levels (since the mass is always constant). With increasing frequency, the damping-dominated region becomes apparent and starting from $f_c/2 \approx 2.4$ kHz, clear TL differences between the various compression levels are evident. It can be observed that the strongest compression (and thus the lowest core layer thickness) result in the highest TL.

The generally higher TL of the test structure at 0.3 d_0 compared to the bare CFRP base structure over the whole frequency range (an increase of 6 dB at 0.5 kHz to 4 dB at 4 kHz) is a result of the approximately 30% mass increase due to the CCLD treatment and the higher damping of the more complex sandwich-like structure. At 4 kHz ($f_c/2 < 4$ kHz $< f_c$) the TL tends to decrease, which indicates that the TL becomes dominated by the damping.

A further benchmark for comparing the measurement results of the different core layer thicknesses is the weighted apparent sound reduction index $R_w{}'$. With this index, the measured frequency-dependent TL can be expressed by a single number. Using the procedure defined in ISO 717-1 [29], the measured TL spectra are shifted and the values of the reference curve at 0.5 kHz are determined. The results for the weighted apparent sound reduction index according to [29] are shown in Table 3.

Table 3. Weighted apparent sound reduction index ($R_w{}'$) of the test structure at various core layer thicknesses compared to the bare CFRF base structure. The values for $R_w{}'$ are determined with the stepwise adjustment (0.1 dB steps) specified in [29].

0.9 d_0	**0.7 d_0**	**0.5 d_0**	**0.3 d_0**	**Base Structure**
28.9 dB	29.4 dB	29.8 dB	30.1 dB	26.1 dB

While the differences between the various core layer thicknesses are rather obvious in the TL spectra above 1 kHz (Figure 5), the weighted apparent sound reduction index (Table 3) is hardly affected by the core layer thickness. As can be seen from Figure 5, at 0.5 kHz, i.e., the frequency at which the weighted apparent sound reduction index is obtained from the reference curve, the TL spectra are hardly affected by the core layer thickness. Consequently, the practical effect on the weighted apparent sound reduction index, which is commonly used in building acoustics, is small. Nevertheless, by a pressure-induced thickness reduction of the core layer, a significant increase of the TL can be achieved above 1 kHz. It should also be noted that the above-mentioned standard is a simplified procedure to determine a single value at 0.5 kHz, but the stiffness, mass and damping dominated regions are strongly dependent on the materials used and the dimensions of the specific CCLD set-up. Thus, with other configurations where the resonance frequency and critical frequency are lower, a stronger improvement of the weighted apparent sound reduction index may well be achieved.

5. Conclusions

The obtained results reveal that the acoustic properties of a structure could be altered using a CCLD treatment. In this publication, the focus was set on the TL, which can be increased using CCLD compared to the CFRP base structure without damping measures. The thickness variation of the core layer (from 0.9 d_0 to 0.3 d_0) set by different actuation pressures and the resulting stiffness and damping changes of the overall structure have no influence on the TL up to approximately 1 kHz for the analysed CCLD design. Between 1 kHz and 5 kHz (upper limit of the measured frequency range), however, it can be observed that the TL can be influenced considerably well by the actuation pressure applied, due to the fact that the TL is damping-dominated around the critical frequency and above. To further investigate the above-mentioned assumptions, measurements in a higher frequency range will be performed. This will also include measurements of the structural damping. In addition, a numerical model should also be built to further validate the measurements and to study the effect of other parameters of the CCLD design, such as thickness ratios or different viscoelastic materials. Finally, it should be noted that the differences between the TL of the various core layer thicknesses are rather obvious in the TL spectra. When assessed according to building acoustics standards using the weighted apparent sound reduction index, no practically relevant effects on the acoustic properties are observed with the analysed CCLD design, since the weighted apparent sound reduction index is obtained at 0.5 kHz, where the effect of the core layer thickness on the TL is small.

Author Contributions: Conceptualization, T.E. and M.D.; Data curation, T.E., M.D. and C.A.; Formal analysis, C.A.; Funding acquisition, N.M. and P.K.; Investigation, T.E. and R.L.; Methodology, T.E.; Project administration, N.M. and P.K.; Supervision, M.D. and P.K.; Validation, T.E., M.D. and C.A.; Visualization, T.E.; Writing—original draft, T.E.; Writing—review & editing, T.E., M.D., R.L., C.A., N.M. and P.K. All authors have read and agreed to the published version of the manuscript.

Funding: The financial support by the Deutsche Forschungsgemeinschaft (DFG, German Research Foundation) project number—315011510 is gratefully acknowledged. The presented work is part of the research within the context of the Priority Program (SPP 1897) "Calm, Smooth and Smart—Novel Approaches for Influencing

Vibrations by Means of Deliberately Introduced Dissipation", subproject "Complex-Shaped Lightweight Structures with Adaptive Dynamic Behaviour through Evanescent Morphing".

Acknowledgments: We acknowledge the Open Access Funding by the Publication Fund of the TU Dresden.

Conflicts of Interest: The authors declare no conflict of interest. The funders had no role in the design of the study; in the collection, analyses, or interpretation of data; in the writing of the manuscript, or in the decision to publish the results.

References

1. Hufenbach, W. *Textile Verbundbauweisen und Fertigungstechnologien für Leichtbaustrukturen des Maschinen-und Fahrzeugbaus: SPP 1123*; Sächsisches Druck-und Verlagshaus: Dresden, Germany, 2007.
2. Kostka, P.; Holeczek, K.; Filippatos, A.; Hufenbach, W. Integration of health monitoring system for composite rotors. In Proceedings of the 18th International Conference on Composite Materials (ICCM18), Jeju, Korea, 21–26 August 2011.
3. Kostka, P.; Holeczek, K.; Hufenbach, W. Structure-integrated Active Damping System: Integral Strain-based Design Strategy for the Optimal Placement of Functional Elements. *Int. J. Compos. Mater.* **2013**, *3*, 53–58. [CrossRef]
4. Adams, R.D.; Bacon, D.G.C. Effect of fibre orientation and laminate geometry on the dynamic properties of CFRP. *J. Compos. Mater.* **1973**, *7*, 402–428. [CrossRef]
5. Alam, N.; Asnani, N.T. Vibration and damping analysis of fibre reinforced composite material plates. *J. Compos. Mater.* **1986**, *20*, 2–18. [CrossRef]
6. Dannemann, M.; Täger, O.; Modler, N. Combined semi-analytical and numerical vibro-acoustic design approach for anisotropic fibre-reinforced composite structures. *J. Sound Vib.* **2017**, *404*, 1–14. [CrossRef]
7. Hanselka, H.; Hoffmann, U. Damping characteristics of fibre reinforced polymers. *Tech. Mech.* **1999**, *10*, 91–101.
8. Holeczek, K.; Starke, E.; Winkler, A.; Dannemann, M.; Modler, N. Numerical and experimental characterisation of fibre-reinforced thermoplastic composite structures with embedded piezoelectric sensor-actuator-arrays for ultra-sonic applications. *Appl. Sci.* **2016**, *6*, 55. [CrossRef]
9. Hurlebaus, S.; Gaul, L. Smart structure dynamics. *Mech. Syst. Signal Process.* **2006**, *20*, 255–281. [CrossRef]
10. Marakakis, K.; Tairidis, G.K.; Koutsianitis, P.; Stavroulakis, G.E. Shunt Piezoelectric Systems for Noise and Vibration Control: A Review. *Front. Built Environ.* **2019**, *5*, 64. [CrossRef]
11. Preumont, A. *Vibration Control of Active Structures. An Introduction*, 3rd ed.; Springer-Verlag: Berlin/Heidelberg, Germany, 2011.
12. Rao, M.D.; Echempati, R.; Nadella, S. Dynamic analysis and damping of composite structures embedded with viscoelastic layers. *Compos. B Eng.* **1997**, *28*, 547–554. [CrossRef]
13. Trindade, M.A.; Benjeddou, A. Hybrid Active-Passive Damping Treatments Using Viscoelastic and Piezoelectric Materials: Review and Assessment. *J. Vib. Control* **2002**, *8*, 699–745. [CrossRef]
14. Zeller, P. *Handbuch Fahrzeugakustik. Grundlagen, Auslegung, Berechnung, Versuch*, 1st ed.; Vie-weg+Teubner Verlag: Wiesbaden, Germany, 2009.
15. Ehrig, T.; Holeczek, K.; Kostka, P. Experimental investigations of lightweight structures with fluidically actuated Compressible Constrained Layer Damping. *Mater. Today Commun.* **2018**, *16*, 204–211. [CrossRef]
16. Holeczek, K.; Koschichow, R.; Schlieter, T.; Ehrig, T.; Kostka, P. Numerical investigations of polymer-based fibre-reinforced structures with fluidically actuated Compressible Constrained Layer Damping. *Proc. Appl. Math. Mech.* **2018**, *18*, e201800260. [CrossRef]
17. Ehrig, T.; Holeczek, K.; Modler, N.; Kostka, P. Dynamic Behaviour Adaptation of Lightweight Structures by Compressible Constrained Layer Damping with Embedded Polymeric Foams and Nonwovens. *Appl. Sci.* **2019**, *9*, 3490. [CrossRef]
18. Holeczek, K.; Zhou, B.; Kostka, P. Evanescent morphing for tuning the dynamic behavior of composite lightweight structures: Theoretical assessment. *Mech. Adv. Mater. Struct.* **2019**, *10*, 1–10. [CrossRef]
19. D'Alessandro, V.; Petrone, G.; Franco, F.; Rosa, S. A review of the vibroacoustics of sandwich panels: Models and experiments. *J. Sandw. Struct. Mater.* **2013**, *15*, 541–582. [CrossRef]
20. Isaac, C.W.; Pawelczyk, M.; Wrona, S. Comparative Study of Sound Transmission Losses of Sandwich Composite Double Panel Walls. *Appl. Sci.* **2020**, *10*, 1543. [CrossRef]

21. Dym, C.L.; Lang, D.C. Transmission loss of damped asymmetric sandwich panels with orthotropic cores. *J. Sound Vib.* **1983**, *88*, 299–319. [CrossRef]
22. Shengchun, W.; Zhaoxiang, D.; Weidong, S. Sound transmission loss characteristics of unbounded orthotropic sandwich panels in bending vibration considering transverse shear deformation. *Compos. Struct.* **2010**, *92*, 2885–2889. [CrossRef]
23. Ehrig, T.; Modler, N.; Kostka, P. Compression and frequency dependence of the viscoelastic shear properties of flexible open-cell foams. *Polym. Test.* **2018**, *70*, 151–161. [CrossRef]
24. Hufenbach, W.; Blanchet, D.; Dannemann, M.; Kolbe, F. Integrated Calculation of Transmission Loss of Sandwich Structures using Coupled Finite and Boundary Element Method. In Proceedings of the 1st ESI Global Users Conference and Exhibition on Simulation-Based Design, Munich, Germany, 19–20 May 2010.
25. Liu, Y.; Sebastian, A. Effects of external and gap mean flows on sound transmission through a double-wall sandwich panel. *J. Sound Vib.* **2015**, *344*, 399–415. [CrossRef]
26. Liu, Y. Sound transmission through triple-panel structures lined with poroelastic materials. *J. Sound Vib.* **2015**, *339*, 376–395. [CrossRef]
27. Du, S.; An, F.; Liu, B. On the sound transmission loss of finite plates with constrained viscoelastic layer. *Appl. Acoust.* **2019**, *149*, 32–38. [CrossRef]
28. DIN EN ISO 10140-4:2010. *Acoustics–Laboratory Measurement of Sound Insulation of Building Elements–Part 4: Measurement procedures and requirements*, German version; Beuth Verlag GmbH: Berlin, Germany, 2010.
29. DIN EN ISO 717-1:2013. *Acoustic–Rating of Sound Insulation in Buildings and of Building Elements–Part 1: Airborne Sound Insulation*, German version; Beuth Verlag GmbH: Berlin, Germany, 2013.
30. Fahy, F.J.; Gardonio, P. *Sound and Structural Vibration: Radiation, Transmission and Response*, 2nd ed.; Acadmic Press: Boston, MA, USA, 2007.
31. Cremer, L.; Heckl, M.; Petersson, B.A.T. *Structure-Borne Sound: Structural Vibrations and Sound Radiation at Audio Frequencies*, 3rd ed.; Springer-Verlag: Berlin/Heidelberg, Germany, 2005.
32. Leissa, A.W. *Vibration of Plates*; National Aeronautics and Space Administration: Washington, DC, USA, 1969.
33. Carboplast Faserverbund-Werkstoffe, TDB-CFK_Platte2mmCarboplast, Product Datasheet, Nov. 2009. Available online: https://www.carboplast.de/index.php?controller=attachment&id_attachment=107 (accessed on 31 January 2020).

MDPI
St. Alban-Anlage 66
4052 Basel
Switzerland
Tel. +41 61 683 77 34
Fax +41 61 302 89 18
www.mdpi.com

Materials Editorial Office
E-mail: materials@mdpi.com
www.mdpi.com/journal/materials

www.ingramcontent.com/pod-product-compliance
Lightning Source LLC
Chambersburg PA
CBHW061558210926
51850CB00009B/2139
* 9 7 8 3 0 3 6 5 2 3 8 1 1 *